HISTORICAL NOTES

MATHEMATICS THROUGH THE AGES

COMAR

COMAP

HISTORICAL NOTES

MATHEMATICS THROUGH THE AGES

A project of the
Consortium for Mathematics and Its Applications,
Lexington, Mass.

ISBN 0–912843–24–1

Suite 210
57 Bedford Street
Lexington, MA 02173
(617) 862–7878

Printed in the United States of America.

Table of Contents

FOREWARD

For eight years, *Consortium*, COMAP's high school newsletter, has provided exciting ideas for introducing students to contemporary applications in mathematics. One of the most popular features of *Consortium* has been our "Historical Notes" column containing informative articles on the history of mathematics.

Historical Notes: Mathematics Through the Ages is a collection of all the Historical Notes articles that have appeared in *Consortium* through issue number 42, Summer 1992. These may be photocopied and distributed to students, with no permission needed from the publisher.

Learning the Ropes:
The Origins of Geometry

Joseph W. Dauben

Euclid's great book on geometry, *The Elements*, is easily the best known work in the entire history of mathematics. It has been read by more people than any other scientific treatise, sometimes for pleasure, often for profit, usually in school. Until recently, it was even read in the original Greek by generations of English schoolboys. Long regarded as an indispensable part of any mathematician's training, it is often the first introduction students are given to the concept of mathematical proof. It seems appropriate, therefore, to devote the first article in "Historical Notes" to Euclid and the origins of geometry.

What may come as a surprise to anyone reading *The Elements* for the first time is the number of references Euclid makes to the word *geometry*: exactly none. The original Greek title is simply *The Elements*, and one turns the pages of the book in vain trying to find any reference at all to "geometry." The word does not appear. What one does find is the now familiar progression from definitions, axioms, and postulates to theorems. Some of these, however, are rather odd. For example, Euclid's fourth definition is:

> "A straight line is a line which lies evenly with the points on itself."

What is this supposed to mean?

Mathematicians today would expect the more appropriate (and usual) definition that a straight line is the shortest distance between two points, a definition first advanced, it seems, by Archimedes. Why then should Euclid define a straight line as that which "lies evenly" upon itself? The idea is confusing and not very precise, yet one expects mathematics to be exactly that precise, if nothing else.

The first book of Euclid's *Elements* is devoted to propositions leading up to the most famous of all theorems in geometry, the Pythagorean Theorem. The Pythagorean Theorem refers to right triangles, perhaps best known in its algebraic form as $a^2 + b^2 = c^2$, where c is the "hypotenuse" (as in **Figure 1**). Of all the words in Greek geometry, *hypotenuse* is one of the most interesting and unusual. But does it have any special significance? Other geometric terms, like *point* and *line*, seem familiar; *hypotenuse* is obviously foreign. What did it mean to Greeks? In fact, the answer to this question, as well as to the problem of Euclid's rather strange definition of "straight line," and even the reason why he never used the word "geometry" in his own great work on the subject, may be found in the origins of geometry in ancient Egypt.

The key to answering all of the above questions is actually "geometry" itself. The word, as used in English, goes back to Latin, and even earlier Greek roots. Its ancient meaning may be seen in the words that comprise it, namely the Greek *gea* or Latin *geo*, meaning "earth," and the word *metria* meaning "measure." *Geometry* in Greek, therefore, means literally "earth measurement." But the ancient Greek historian Herodotus tells us that mathematics arose first in Egypt, because of the need to survey land quickly after the annual flooding of the Nile. Other ancient authors, including Heron of Alexandria, Diodorus, Siculus, and Strabo, agree. The Egyptian scribes, sent to measure out parcels of land in order to levy accurate taxes, took *harpedonaptai* with them. As another Greek writer, Democritus, explains, the harpedonaptai were "rope stretchers," a literal translation of the Egyptian name. These were the royal surveyors of Egypt; there is a graphic wall painting in the Tomb of Djeserkere-sonb at Thebes in Egypt that depicts such harpedonaptai with their ropes, ready to survey the land.

Since little more is known about the Egyptian harpedonaptai than their name, it is helpful to turn to other sources that give more details about rope stretchers. In fact, among the Akkadians, Assyrians, Babylonians, Hebrews, and Arabs, surveying was done by rope stretchers. The Talmud, for example, gives exact details in many places about how the rope stretchers were to go about their work. In the Old Testament, Isaiah 35:17 describes land as "portioned out to them with the line," and Amos 7:17 proclaims that "land shall be parceled out by line."

Some of the ancient knowledge proceeding the Greeks was relatively advanced. In India, for example, it was known as early as the eighth century B.C. that a triangle with integer sides of 3–4–5 was a right-angled triangle. Several centuries later, the Apastamba Sulva-Sutra (fifth–fourth century B.C.) described right angles by means of stretched ropes, using such Pythagorean triples as 5–12–13, 8–15–17, and 12–35–37, to construct rectangular altars. In China, knowledge of simple Pythagorean triples was used to produce *gnomons* or the *carpenter's square*, which provided the right angles used by artisans and architects in their daily trades. Otto Neugebauer has even suggested that the Babylonians may have had a formula for generating such Pythagorean triples, considering the large number of such combinations that have been found on cuneiform clay tablets that record much of Babylonian mathematical knowledge.

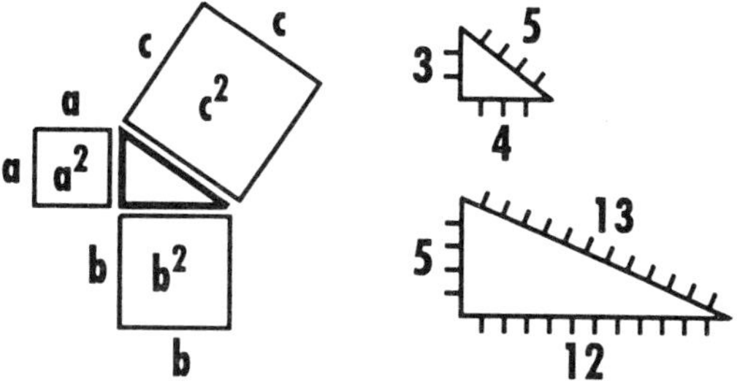

FIGURE 1. THE GEOMETRIC VERSION OF THE PYTHAGOREAN THEOREM: THE SUM OF THE SQUARES OF THE SIDES OF ANY RIGHT-ANGLED TRIANGLE IS EQUAL TO THE SQUARE OF THE HYPOTENUSE.

THE TWO TRIANGLES REPRESENT TWO SPECIFIC EXAMPLES OF SPECIAL CASES OF THE THEOREM, NAMELY THAT FIGURES WITH SIDES OF INTEGER LENGTHS 3–4–5 AND 5–12–13 ARE RIGHT-ANGLED TRIANGLES.

Other Names in Ancient Cultures

Equally of interest, however, is the nomenclature adopted by these ancient cultures in describing their geometrical practice. In Sumerian, for example, the word for line is *tim*, which originally meant "cord" or "rope." Similarly, our English word "line" is derived from the Latin *lines*, which comes from *linum* meaning "cord" or "thread of flax." The oldest Hebrew geometry, described by one commentator as a sort of practical handbook for rope stretchers, is the *Mishnat ha-Middot*, or "Theory of Measures." It also uses the word "cord" or "rope" for "diagonal." Both Chinese and Hindu writers used the word "rope" to describe the diagonal of a square, the hypotenuse of a right isosceles triangle. Finally, the Hindu word for geometry is *Sulva* or *Rajju*. In Sanskrit, both words have the same meaning, that of "rope" or "cord." In ancient India, the geometer was called, literally, a "uniform rope stretcher" or "rope holder."

But what has all of this to do with Greek geometry? Since virtually all ancient commentators agree that Greek mathematics originated in Egypt, does the practice of the Egyptian harpedonaptai shed any light on the peculiarities of Greek geometry that we have already mentioned? Above all, the experience of the rope stretchers was transmitted directly into Greek geometry in the most basic of ways: terminology.

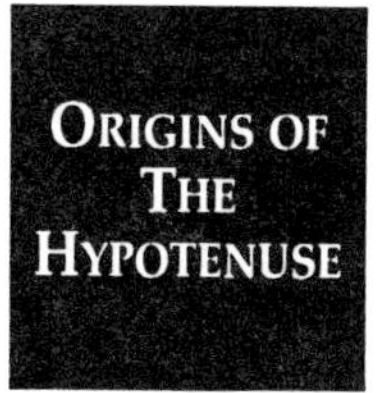

The Egyptian harpedonaptai provide the answer for that most peculiar word, *hypotenuse*. Both words involve the idea of stretching. Hypotenuse is derived from the Greek word *teinousa*, which means "stretched." This is exactly the sense in which Plato describes the diagonal of the square in his dialogue, the *Meno* [48E8–85A2], where he used the word *teinousa* to indicate the line "stretched across" from opposite corners of the figure. This same meaning is conveyed by the Greek word *hypotenuse*—"that which is stretched over" or "across," just as the rope stretcher would pull ropes tight to form the hypotenuse between the sides of a right triangle.

Furthermore, the legacy of land surveying and the practice of the Egyptian rope stretchers survives in Euclid's definition of a straight line. To lie "evenly with the points on itself" is a reflection of what a rope, pulled tight, would have to exhibit in order to lay out a straight line. This is perhaps seen more clearly in the definition given prior to Euclid by Plato. Plato defines a straight line in the *Parmenides* [137E] as "that of which the middle covers the ends." The same idea is also used by Heron to define a straight line as "that stretched to the utmost between the endpoints."

Compare these with the better known definition of Archimedes, that a straight line is the shortest distance between two points. In contrast to Euclid's definition, Archimedes's is elegant and abstract, and emphasizes the basic metric nature of geometry by focusing on the mathematical space between points. It is a definition much better suited to the austere and precise terms of axiomatic mathematics, but it fails to convey the humbler origins of Greek geometry in the practical business of the surveyor.

This may also help to explain why Euclid did not call his great work a treatise on geometry, nor did he use the term anywhere in *The Elements*. Instead, he seems to have succeeded in avoiding any connection between the ideal "elements" of his mathematics and the mundane objects of day-to-day living. Euclid doubtlessly regarded his *Elements* as a lofty, theoretical, axiomatic treatment of an area of mathematics, better known as *geometry*. Applications, including the demeaning labor of field surveying and the word "geometry," with its direct connotations of land measurement, were to be avoided entirely.

Nevertheless, in this use of such terms as *hypotenuse*, and in definitions like that for straight line, the humbler, more practical and empirical origins of geometry were betrayed. The activity of Egyptian rope stretchers provided not only a part of the metaphorical language, but, more importantly, the initial conceptualization upon which the Greeks constructed their impressive series of theorems and proofs, culiminating in Euclid's elegant, axiomatic geometry. ❑

Indian Rope Tricks:
A Number of Knotty Problems

JOSEPH W. DAUBEN

In 1680, the Pueblo Indians of Arizona coordinated a revolt against the Spanish with literally a string of messages. Cords were knotted and distributed among the indians. By untying one knot each day, the countdown terminated in a carefully synchronized uprising. Other American Indian tribes also counted by means of knots on string, often using different colors to differentiate and keep track of time, especially days, moons, and winters. They also fixed the dates of ceremonial observations and hostile raids in a like manner.

American Indians were not the only ones to see the value of counting by knots. The Greek historian Herodotus tells the story of Darius, the Persian king who drove his army across the Ister river in pursuit of the Scythians. Posting a guard at a strategic bridge spanning the river, Darius left a rope tied with 60 knots to be untied, one each day. If he failed to return by the time all of the knots were gone, those left to guard the bridge were instructed to sail for home.

Earlier still, in the great Chinese classic, the *I Ching*, reference is made to knotted cords used "for the administration of affairs." Tradition dates the use of such cords from about 2800 B.C. For more than two millennia, they served basic record-keeping functions, until they were superceded by written characters, probably during the sixth century B.C.

Lao-tzu (551–479 B.C.), author of the famous *Tao-te Ching*, even called for a "return to the spirit of the olden days when they used knotted cords for their records." Lao-tse actually elaborated a system of knot-counting in terms of the concepts of Yin-Yang (the harmony of opposites, or the *T'ai Ch'ai*). Odd numbers were counted by white knots (representing that which was complete, as day, warmth, and the sun), and even numbers by black knots (representing that which was incomplete, as night, the cold, water, and earth).

The use of knots tied in a string to record statistics also has a counterpart that persists to the present day. In the last issue of *Consortium*, the use of ropes as an instrument of land surveying was explored, especially as it was related to the origins of geometry and the Greek word, *hypotenuse*.

Many ancient cultures used ropes to expedite land measurement. A similar idea, using a line knotted at equal intervals, makes it possible to measure not only lengths, but speed as well, in a way that is familiar to seafarers and Sunday sailors alike. The measure of nautical speed, the "knot," is literally a measure of knots. The mariner's log line consists of a rope with knots tied at equal intervals (the length turns out to be 47 feet, 3 inches). The first length ends with a single knot, the second with two knots, the third with three knots, etc. When thrown out astern, it is possible to determine the speed of the vessel in nautical miles per hour—or knots—by counting the number of knots passing through the mariner's hands in 28 seconds. (Traditionally, this was the interval used, the time being measured by a small glass similar to an hourglass or modern egg timer.)

The most remarkable example, however, of knot records in the annals of history is that of the *quipu* (pronounced *key-poo*) of the pre-Columbian Incas. Despite a highly evolved and complex civilization, with an elaborate government bureaucracy and a spoken language (*Quechua*), the Incas seem never to have evolved a system of writing, neither heiroglyphic nor phoenetic. Thus, the matter of record keeping was carried out in an alternative, yet very practical, way by means of knots on ropes. In fact, the word *quipu* simply means "knot" in Quechua.

A quipu is nothing more than a cord with knotted strings attached to it, each knot representing a unit or decimal group of tens, hundreds, and in special cases, even thousands. By using different colors, grades, and lengths of string on the same quipu, distinctions could be made between different groups or types of objects being counted. By tying knots at uniform intervals and in groups, gaps could be used to indicate zeros. Clearly, the Incan quipu was a decimal, place-holding number system of considerable flexibility.

Quipus were used to maintain records of all sorts, especially for the government. Those responsible for knotting and interpreting the quipus were the *Quipucamayoc*. These members of the Incan society were given special training and assumed a position of considerable importance. The most important quipu records were undoubtedly census data, tax lists, storehouse inventories, land surveys, property holdings, and the like. Many of the early chroniclers indicate that there were also quipus that served to preserve non-numerical information, including Incan history, especially the chronology of kings, as well as traditions, folklore, and even poems. However, most experts regard such reports as exaggerated, indicating at best some sort of mnemonic function whereby knots were used to aid memory, like rosaries aid the devout in remembering the order of certain prayers. The knots or beads, however, do not provide a literally encoded account of information in any narrative sense.

Of all the suggestions made about possible uses of the quipu, one of the most interesting is due to the Baron Erland E. Nordenskiöld, a notable and early student of the quipu. Nordenskiöld maintains that some quipus served as calendars, recording the solar year as well as the synodic revolutions of the planets, especially Venus (as was known in Mayan codices), and perhaps even Mercury. Other writers have argued that quipus may also have been used to predict lunar eclipses, and that Incan astrologers fixed dates for sowing, harvesting, and celebrating the festivals of the sun by using the quipu calendar records. In fact, the Aztecs of Mexico, known for their astronomical calendars, used a special word for periods of 52 years, *xiuh-molpilli*, which means "knotted years."

FIGURE 1. A QUIPUCAMAYOC WITH AN UNKNOTTED QUIPU.

Of major significance in documenting the use of quipus by the Incas is a series of remarkable illustrations from a manuscript (totaling 1129 pages) written by Felipe Guamán Poma de Ayala for the King of Spain (probably compiled between 1595 and 1613). Known as the *Nueva Corónica y Buen Gobierno*, it was forgotten for centuries but rediscovered in 1908 in the Royal Danish Library in Copenhagen, Denmark. The document includes 397 drawings, a number of which depict officials with quipus. The one reproduced here (in **Figure 1**), in fact, shows an official holding an unknotted quipu. The top line, in Spanish, indicates that this is indeed a *Cōtador*, meaning "counter" or "one who counts"; the next lines give the Quechua equivalent.

What is particularly interesting about this illustration is the combination of the Quipucamayoc with his quipu stretched out, and the device at his side, which indicates some form of counting board or abacus. This probably used lima beans or kernels of maize as counters. The device would have enabled the Quipucamayoc to carry out arithmetic calculations based upon data drawn from his quipus. No doubt, the results would then have been recorded, either on the same quipu at an appropriate place, or on another quipu.

This illustration, however, makes clear the major drawback of the quipus. Although they were a convenient, useful means of storing information, they could not have served the Incas as practical calculators. For this, something very much like the abacus was necessary. But the two together, quipu and abacus, were obviously of great service: the one for carrying out temporary calculations of the moment, the other for preserving results in a relatively simple, compact, permanent form.

The word "abacus," in fact, may betray an intimate connection with the quipu-like records. If it is derived from the ancient word *abac* in Hebrew-Aramaic, possible meanings include "loop" or "knot." This would suggest that the Old World abacus possibly may have had its beginnings in the idea of record keeping through knotted cords. ❑

"The King was in His Counting House..."

Joseph W. Dauben

The Museo Nazionale in Naples, Italy, contains many wonderful objects from antiquity. Not the least of these is an imposing Attic red-figured amphora of considerable size, richly decorated with scenes depicting the Persian Wars. At the center is the figure of Darius, King of the Persians, being warned of the consequences of his planned campaign against the Greeks.

Figure 1. The Royal Treasurer of Persia, shown at his counting table, tablet in hand. This drawing of figures on the Darius Vase in the Museo Nazionale in Naples, Italy, is reproduced from Karl Menninger's book, *Zahlwort und Ziffer* [1934].

For the history of mathematics, however, the Darius vase is particularly significant because of one small vignette just below the central figure. This vignette depicts the royal treasurer who is keeping the accounts of tribute paid by subjects to the king. (See **Figure 1.**) What is of interest is the graphic representation of how the numbers were calculated and recorded: tributes were entered on a counting board, added by appropriate combinations of moveable markers, and then recorded permanently on a hinged pair of bronze slates.

Darius was mentioned in the last issue of *Consortium*, in connection with knotted ropes used as record-keeping devices. The knots, each representing a unit of one day, were taken together to signify the total number of days a garrison of soldiers, sentried at a strategic point, were to remain on duty. An even more advanced form of such record keeping on knotted cords, used in concert with an abacus-like device for calculations, was also discussed in that previous article.

The *quipu*, a complex combination of cords and knotted strings, was used with great skill by the pre-Columbian Incas. A number of these quipus have been preserved, and the knot-tiers (or *Quipucamayocs*) have even been illustrated in a manuscript known as the *Nueva Corónica y Buen Gobierno* by Felipe Guamán Poma de Ayala, sometime between 1595 and 1613.

One of these illustrations shows a Quipucamayoc with a counting board and counters at his side. The counting board—looking very much like a large checker board—would have facilitated arithmetic calculations, perhaps using data drawn from the quipu; the results could then have been permanently recorded, again on the quipu. The Darius vase, however, represents an even more advanced use of mathematics than that of the Quipucamayocs, for it shows the combination of an actual device enabling easy calculations with a written means of recording those results in a permanent fashion.

The Darius vase is by no means the earliest record we have of such devices. There is a remarkable piece of white marble preserved in the epigraphical collection of the National Museum in Athens. It was found on the Greek island of Salamis, and represents the earliest physical evidence yet known of a device meant to simplify and accelerate arithmetic calculations.

Although it is not possible to give the Salamis tablet a precise date, the Greek number-letters inscribed on the marble itself attest to its antiquity. Already by the seventh century B.C., the use of such counting boards must have been generally familiar, for the Greek law-giver Solon (according to Diogenes Laertius) made a political reference to those who had influence with tyrants as being like pebbles on a counting board, their values standing sometimes for more and sometimes for less.

By the second century B.C., this same allusion was still popular, as the Roman author Polybius attests:

"These men [who surround the tyrant or king] are really like the pebbles on a counting board. The pebbles, according to the pleasure of the reckoner, may be valued at one moment as little as a chalkous, or at the next moment as much as a talent."

The Greeks called a counting board an *abax* and the counters *psephoi*, or "pebbles." Nevertheless, there is considerable controversy as to the actual origin of the word *abacus* itself. The simplest, if least convincing suggestion, is that made by Johannes de Muris (ca. 1350), who held that the word was taken from the Hebrew *abaq* meaning "dust," and refers to sand or dust boards on which lines could be traced with a finger or other instrument. Calculations could thereby be made quite easily, with erasures and reformattings a simple matter, especially if working in sand. Still others have maintained that *abaq* can also mean "loop" or "knot" in Hebrew-Aramaic, and that the word *abacus* may be related to knotted cords once used for arithmetic record keeping.

This view draws some support from the fact that in Arabic, tens and hundreds are referred to as *cuqud*, meaning "knots," and that the Arabic mathematician Al-Khwarizmi said that numbers consist of three essential types: units, knots (or tens), and composites. Whatever the origin of the Greek word *abax/abakos*, however, it is clear that in the Greek mind, the basic notion of arithmetic was related to the use of stones. One form of the Greek verb "to count" was literally "to pebble," which attests to the early association of counting with pebble arithmetic. Actual arithmetic calculations would have been further simplified by using graduated or place-valued counting boards.

In Latin, the word *abacus* was borrowed from the Greek *abakos*. The Romans translated directly to name their counters *calculi* (or *abaculi*), which is the plural of *calculus*, meaning "little stone" or "pebble." (See **Figure 2.**) Our words calculate, calculation, or calculator are all derived from this sense of the ancient way of handling simple arithmetic combinations through use of moveable calculi. In fact, the Romans expressed the procedures of calculating as "putting" or "removing" pebbles.

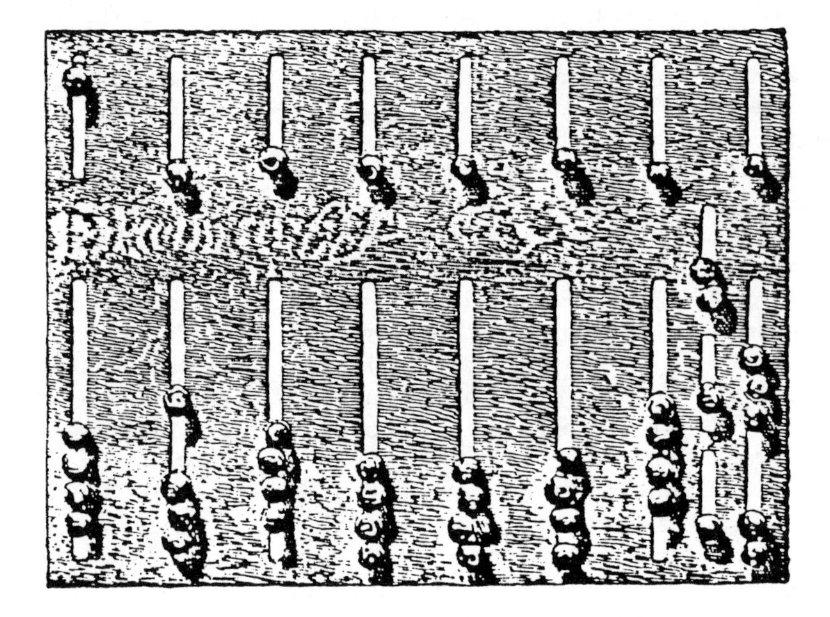

FIGURE 2. EXAMPLE OF A ROMAN ABACUS CONSISTING OF GROOVES AND MOVABLE COUNTERS OR *CALCULI*. THIS DRAWING OF A BRONZE ABACUS, NOW IN THE BRITISH MUSEUM IN LONDON IS FROM D.E. SMITH'S BOOK, *HISTORY OF MATHEMATICS*, VOLUME II [1925].

Latin has another word, however, for the individual who actually performed such calculations on a counting board, namely *mensarius*, which means "tabler"—literally, one who uses the table or counting board. In the Middle Ages, it was only a short step to associating those who were in the business of using such tables for changing or lending money, namely those who used counting boards or counting "banks"—with a new word, "banker." Nevertheless, the older meaning of table survives to this day in yet another English word, "banquet" (actually borrowed from French), referring to a sumptuous meal laid out on long tables. Bankers, however, used such tables with incised lines to carry out their various arithmetic calculations.

The French Influence

The mathematical experience of the medieval banker has an especially informative history in France, and one with direct links to numerous words and concepts that are familiar in English today. For example, the use of counting boards was described in 1474 in the *Mémoirs* of Olivier de la March. It seems that Charles the Bold of Burgundy liked to carry out computations of his own at the state treasury. Often he could be found there, at the counting board (using his own counters of gold), keeping track of the treasury, and an eye on his treasurers, too, no doubt. What he counted on was a *bureau*, one of the words in French corresponding to board or counter in English. The word "bureau" has been carried over not only into modern French and German to mean the actual office where such administrative record keeping is performed, but into English where "bureau" may even refer to large administrative offices, like the Federal Bureau of Investigation.

An equally interesting history is associated with the English word "Exchequer." This goes back to the twelfth century when Henry II was king, not only of England, but of a good part of France as well. It was then that the English borrowed their term for the treasury from the French word *eschequier* (in medieval days, *echiquier* in modern French) doubtless because the practice of banking was more highly developed in France at that time.

Counting and Chess

The word *echiquier* itself brings us again to the title of this article, because the word traces its roots to *scaccarium* in Latin, which means "chess board," (as in **Figure 3**) a literal reference to the checkered pattern of lines on counting boards, resembling the divisions of a chess board from antiquity to the present.

In fact, the Latin word for such a checkered pattern goes back to the game of chess, which has its origins in ancient Persia. The dominant marker on the chess board is the king—*shah* in Persian. Thus, we move from the Persian *shah* to the French *echec* for chess (in German, chess is *Schach*, an even closer relative of *shah*), whence *echiquier*, and finally to the English department of revenue, the treasury or Exchequer.

But there is more. We are all familiar with bank checks and the ease of handling many financial transactions by check. This word, too, goes back to the mathematical practice of medieval bankers, and especially to the English

Exchequer, who kept track of income and expenses on bits of notched wood known as *tally sticks*. Such sticks were issued as actual payment for debts. Usually, a stick was marked, then split in half, providing a record for both parties. To redeem the amount paid by the Exchequer on a tally stick, the payee simply took his half to the Exchequer, where the two tally stick halves were matched, or checked. If they agreed, payment was made, and hence the origin of our words "to check," meaning to compare (usually for accuracy). The noun "check," means "receipt" or "written promise to pay." The word "tally" has had a similar fate in English: to tally meaning "to add up, compare, or agree," and tally (the noun) meaning "an addition" or "sum."

Finally, our notion of bankruptcy is derived from this same constellation of ideas, and again takes us back to medieval counting boards. Italians, for example, use the colorful expression *banca rota*, which means "broken bank or board." Our word "bank," in fact, comes directly from the words for board, banque, or bank. The idea of broken board for bankruptcy is a direct reference to the practice of breaking the counting boards, the banks, of money changers inclined to cheat. In English, "bankruptcy" carries this same meaning in terms of the rupture of the bank, the breaking of the counting board.

The word *echiquier* itself traces its roots to *scaccarium* in Latin, which means "chess board," a literal reference to the checkered pattern of lines on counting boards, resembling the divisions of a chess board from antiquity to the present.

FIGURE 3. THE CHESS BOARD.

The Pharoah and His Fingers

The abacus and counting boards of antiquity and the Middle Ages, however, were not the only devices used to promote a facility with arithmetic. Consider, for example, another king, not in his counting house this time, but one who used his knowledge of numbers to a very different end.

This king is described in the Egyptian Book of the Dead, where the story is told of the Pharaoh who came to the river that must be crossed in order to attain eternal life. There was a ferryman prepared to convey the Pharaoh across the river, but he doubted his identity and asked, "Did you bring me a man who cannot number his fingers?" As a sign that he was in fact a Pharaoh who possessed this difficult power, he proceeded to count his fingers, thanks to a mnemonic device that the Pharaoh had learned. The king could indeed count, but the ability to do so in ancient Egypt was considered arcane, even magical, knowledge in terms of the powers required to do so.

This example represents counting on one of its most immediate levels, the fingers of the hands. This simple device, in fact, has been of momentous importance in the history of mathematics, but the matter of counting on the fingers of one or both hands (and sometimes on the toes of the feet as well) is another story. The human hand, after all represents the first example of a digital computer. ❑

The First Digital Computer:

A Handy Guide to Simple Counting

JOSEPH W. DAUBEN

Ovid, the Roman poet, once wrote that the number ten was always held in high esteem because *seu quia tot digiti*, it is "the number of fingers by which we count." The hand, in fact, is a significant mathematical instrument, and has been used throughout the ages as a means not only for counting but for calculating as well.

Beginning with one and counting the fingers of both hands, it is a straightforward matter of proceeding quite naturally from one to ten, and there is no mystery about the naturalness of counting thereafter in groups of ten. Our word "digit" for "number" is a direct legacy of this method of counting, *digitus* being the Latin word for "finger." Thus, the numbers were literally the fingers (or digits) with which they were associated. Similarly, magicians are sometimes referred to as "prestidigitators," a reference to the "fast fingers" required for true sleight-of-hand.

Digit and finger are also related botanically in the common foxglove, known as well by its Latin name *digitalis* (of the finger). This name was coined by Leonard Fuchs in his herbal of 1542 for the plant's thimble-like flowers; in fact, in German it is called *fingerhut*, likewise meaning "finger cover" or "thimble."

While the Romans associated digits with units, they also gave a special name to groups of ten. The naturalness of counting on the ten digits of both hands led to this special word, namely *articulus*, which means "knot" in Latin. But what does the number ten have to do with *articuli* or knots?

The answer, again, goes back to the hand, and to the Roman way of counting in terms of digits and articles. Originally, *articuli* were "knots that join and bind the bones of the fingers," namely the joints of the hand. Therefore, by counting on digits and articles, the Romans were counting on fingers and joints.

Among the most prominent of the late Roman authors to write about manual arithmetic was Boethius (sixth century A.D.), who divided all numbers into three classes: digits, articles, and *numeri compositi*. The "composite numbers" were simply all those numbers formed from combinations of fingers and joints, or units and tens. Some idea of how this system of manual arithmetic worked is given in a work by the Venerable Bede (A.D. 673–735), the English Benedictine from Northumbria, who instructs as follows in his *De temporum ratione* (A.D. 725):

When you say One, bend the little finger of the left hand, and place it on the middle of the palm.

When you say Two, bend the fourth finger and place it likewise.

When you say Three, do the same with the third finger ....

When you say Ten, put the tip of the first finger on the middle joint of the thumb.

[Quoted from J. Murdoch. 1984. *Album of Science: Antiquity and the Middle Ages*. New York: Scribner's, 79.]

This habit of counting on fingers and joints was noted by the English mathematician Robert Recorde in his work, *The Grovnd of Artes*, as late at 1543, when he wrote of both "digits and articles" in referring to units and tens. The method of counting units of ten on the articles or joints of the hand are seen clearly in **Figure 1**, a woodcut illustrating the *Summa de Arithmetica* of the Italian mathematician Luca Pacioli, published in 1494.

In addition to simple counting, the hand was also used to compute addition and multiplication and to provide solutions for some very intricate problems. The Venerable Bede was especially interested in the question of determining the date of Easter, which was set annually on the first Sunday following the first full moon after the Spring equinox. The complexities of this problem are considerable, because calculating the proper date each year involves the daily rotation of the earth, the lunar monthly period, and the solar year, all of which are incommensurable mathematically, giving rise to great difficulties. [To see a full calculation of the Easter date, see page 84.]

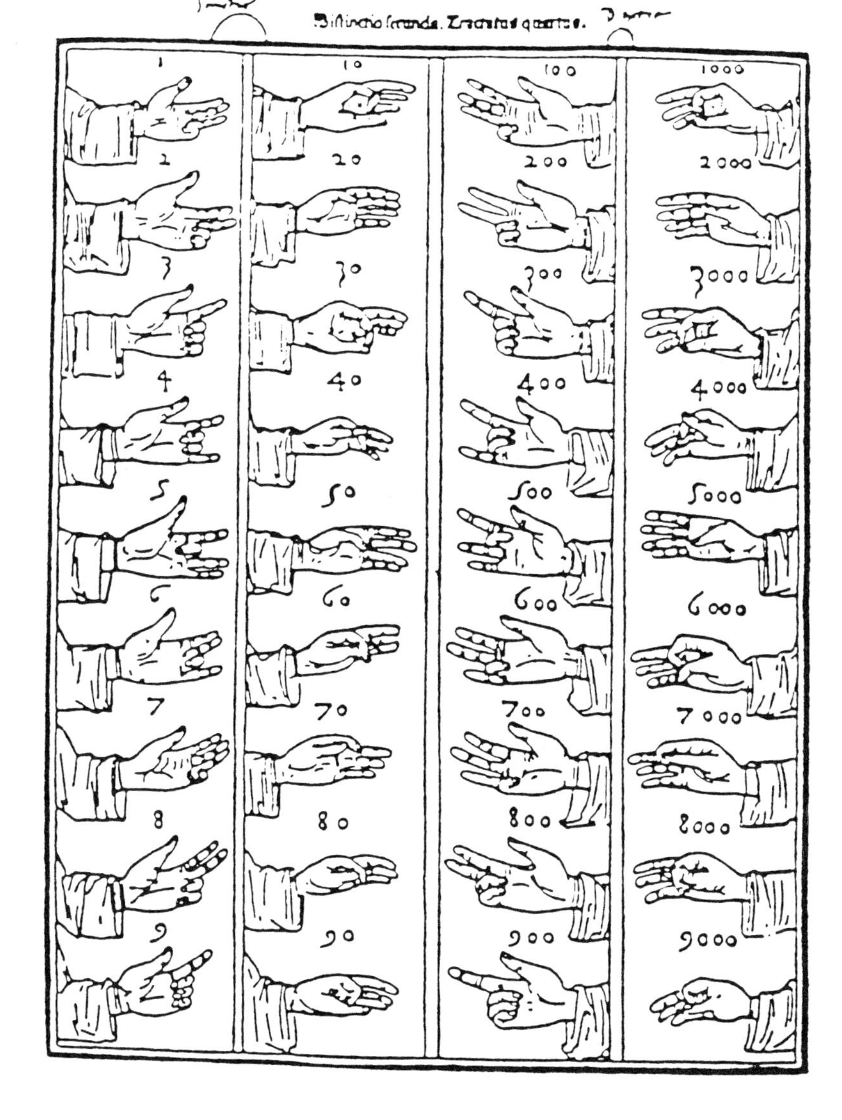

FIGURE 1. THIS GRAPHIC REPRESENTATION OF FINGER COUNTING IS TAKEN FROM LUCA PACIOLI'S *SUMMA DE ARITHMETICA* OF 1494. IT WAS THE FIRST IMPORTANT WORK IN MATHEMATICS TO BE PUBLISHED, AND FIRST APPEARED IN VENICE. THE SECOND COLUMN CLEARLY SHOWS THE COUNTING OF THE ARTICLES FROM TEN TO NINETY ON THE JOINTS OF THE LEFT HAND.

Bede developed a method whereby the calculations required for setting the date for Easter (as well as other moveable feasts of the Church calendar) could actually be carried out using the hand as a computing device. In his *De temporum ratione*, he set out a clear method for calculating Easter tables, and discussed general aspects of astronomical and arithmetic computation as well. It remained for many centuries one of the best expositions of the principles of the underlying construction and calculation of the Christian calendar. Both digits and articles were used to aid the computations Bede describes.

Medieval scholars developed the hand computer to a high art. As the thirteenth century author Leonardo of Pisa (also known as Fibonacci) wrote:

> *Multiplication with the fingers must be practiced constantly, so that the mind, like the hands, becomes more adept at adding and multiplying various numbers.*

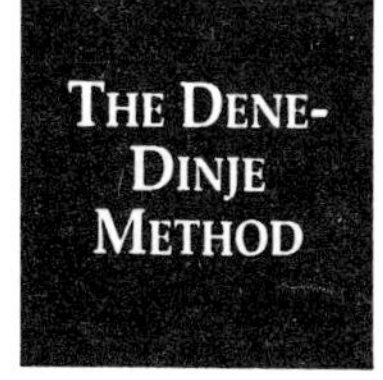

The Romans and their medieval scholastic followers were not the only people to consider the ease of counting manually. The universality of idea is reflected in a method used by the Dene-Dinje tribe of American Indians. The Dene-Dinje counted (and kept track of different amounts) by bending their fingers. Their expressions for this form of counting are equally picturesque, and recall on a more limited level of European finger counting:

one: "the end is bent" (meaning the little finger is bent);

two: "it is bent once more" (now the fourth finger is bent);

three: "the middle is bent" (the middle finger is bent);

four: "only one is left" (the first finger is bent);

five: "my hand is finished" (all five fingers are at last bent; they have all been counted.

By counting on one hand only, grouping by fives is natural, and this became the basic unit of counting among the natives of Enggano, an island just south of Sumatra. There, the word for five is *ariba*, which means "hand," and refers, obviously by now, to the five fingers. Similarly, the Enggano word for twenty is *kahaii ekaka*, which means "one man," again referring to all the fingers and toes together.

But now the counting gets interesting. *Ariba ekaka* (hand man) is Enggano for one hundred, literally five twenties, or a hand's worth (five) of men (twenty) meaning fingers and toes. *Kahaii edndodoka* (one our body) translates as four hundred, and may be interpreted as twenty times twenty in terms of one body (twenty fingers and toes) for each part of one body (twenty fingers and toes again).

In New Guinea, the number 98 would be described in a similar fashion: four men, two hands, one foot and three. In other words, 4(20) + 2(5) + 1(5) + 3.

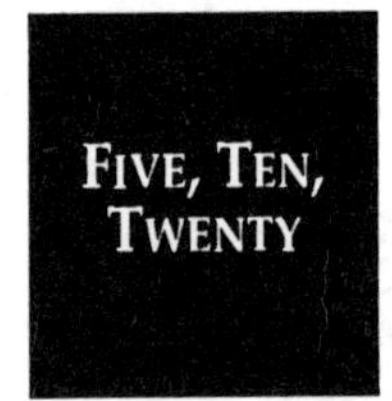

My favorite example, however, of an English word derived from the number five is "punch." The English imported the word, along with the refreshing drink that came to bear the same name, from the East Indies in the seventeenth century. The punch was made from five ingredients, the word having been taken from the Sanskrit-Hindi word *panca*, meaning "five."

There is no reason, however, to limit counting to the hands, either one or two, and other cultures offer interesting examples of number systems based upon both hands and feet, as well as the entire body. Like the Engganos, in Eskimo the number twenty is the word for man. Forty would be two men, and so on.

Obviously, a vestige of such counting by twenties is our word "score," immortalized by Abraham Lincoln in the opening line of the Gettysbury Address: "Four score and seven years...," meaning 87 years. The French language still uses such *vigesimal* (base twenty) counting in expressing numbers between 80 and 100. Eighty is *quatre-vingts* (four twenties); 91 is *quatre-vingt-onze* (four twenties and eleven). In Medieval French, one can find examples in the eleventh century of 60 expressed as *trois-vingt* (three twenties), 120 as *six-vingt* (six twenties), and even 360 as *dix-huit-vingt* (eighteen twenties). Similar versions of vigesimal counting, making good use of numbers whose origins must go back to fingers and toes together, are well-known among the pre-Columbian Mayan and Aztec languages, as well as in Celtic languages like Irish, Scottish, Gaelic, Cornish, and Breton.

Not all cultures, however, have needed to count as far as five or ten. By going no further than two (called a *binary system*), relatively sophisticated numbering is possible. Indeed, there are tribes in Africa, like American Indian counterparts, that count only as far as two—anything beyond that is simply "many" as in "one, two, many." There is even a group of South Sea Islanders that counts in a roughly binary fashion, using only the words *urapun* for one and *okasa* for two. These are the only numbers they have, and yet they do not stop at that with a simple "many" to cover anything greater than *okasa*. Instead, they have provided names for numbers running successively higher up the number sequence by combining their two concepts of *urapun* and *okasa*.

Three, for example, is *okasa urapun;* four is *okasa okasa.* What would five be? Twenty-five? In trying to handle even so simple a number as ten, *okasa okasa okasa okasa okasa,* it is obvious that such a binary system based only on ones and twos is too cumbersome to facilitate much in the way of arithmetic. This example shows, however, that even with a system limited to only two numbers, one and two, it is still possible to count higher numbers without stopping at "many."

English "Twoness"

The English language offers some telling examples about counting by twos, and a lesson at the same time about the nature of counting, and perhaps even its origins as well. We take the business of counting well beyond two for granted, and yet the abstract ideal of "twoness," as distinct from the number two, is rather remarkable in itself. For example, consider the different ways in which two is expressed when applied to different things like shoes, oxen, or quail. English refers to a *pair* of shoes, a *yolk* of oxen, a *brace* of quail. But we do not speak of a brace of shoes or a yolk of quail. Why?

This is because originally there was no abstract concept of twoness; two quail and two oxen were regarded as different things. Therefore, different words were used to describe their most common pairings. Moreover, the abstract concept of twoness is not so straightforward as it may at first appear to our minds, which are used to dealing in abstractions. We speak of two books, but neither book by itself has the property of twoness. No two of anything has this property by themselves. Clearly, the advance to an abstract concept of number, where two could be applied to anything coupled in pairs (whether shoes, oxen, or quail) was a significant advance conceptually.

Other Words Based on Two

Counting by twos has also served to enrich English with many words which, derived from Latin or other sources, often conceal their original meanings. The words "diploma" and "diplomat" offer but two very interesting related examples. Both words hinge literally on the number two, taken from the Greek *di-ploos* or the Latin *duplus. Duplus* eventually became "double" in English.

The story of "diploma" and "diplomat" goes back to the custom of Roman emperors who granted certain privileges by writing decrees on bronze plates, which were joined by two rings and then sealed with wax. The two plates were called a *diploma* (two tables), which constituted an imperial document stating the privileges granted. A *diplomat,* therefore, was one who carried such diplomas or doubled tables to foreign governments on official visits. Universities today grant diplomas to their graduates, which describe the "rights and privileges" to which they have been admitted as members of the learned community of scholars, following successful completion of their courses of academic study.

On the other hand, there are many words that are based upon another aspect of the number two, namely the division or separation implied by an expression like "divided in two." "Divide," in fact, conceals the word "two" twice, since *di–* comes from the Latin prefix *dis–*, meaning "to separate into two equal parts"; and the prefix *vi–*, also meaning "two," as in the Latin *vi–ginti,* meaning "twenty," from which words like *vigesimal, vignt* in French, or *veinte* in Spanish are derived.

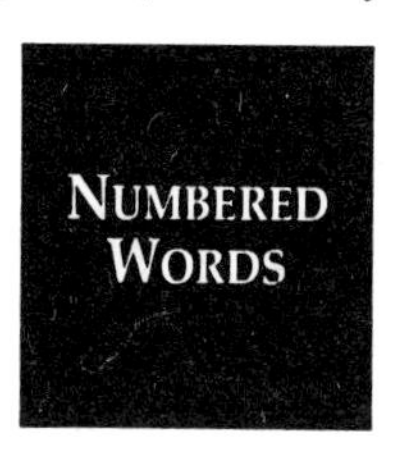

Numbered Words

Today, counting is such a natural part of everyday life that we may take it for granted without realizing how often we are using words and concepts that refer in one way or another to numbers or counting. There are numerous examples of names given to objects by virtue of how we count. In music, groups are commonly called trios, quartets, quintets, sextets, or octets. Years may be counted in decades (literally by tens), while athletes may participate in decathlons. *Decathlon* arises from the prefix *deca–* meaning "ten" and the Greek word *athlon,* meaning "contest"; a decathlon is a contest involving ten different sporting events (just as a *pentathlon* involves five events).

We are also used to counting years by decades, making change for a dollar with dimes, and dividing the year into months, ending with December, clearly a name in which the Latin word for ten, *decem,* is nicely imbedded. All of these names, in fact, are based in one way or another on decimal systems, anything based on or proceeding by units (or divisions) of ten.

December, however, seems wrong somehow. It would make sense if there were ten months, but instead there are twelve. Why twelve? And why is the last month, December, named the tenth month if it is the last of twelve? Shouldn't it be "Duodecember," or the twelfth month? Answers to these questions will be the focus of the next Historical Notes column. ❑

March of the Months
(Mathematically Speaking)

JOSEPH W. DAUBEN

As I am sitting at my desk drafting this column, it is mid-March and I am contemplating the months of the year and the numerical discrepancy noted in an earlier issue of *Consortium*, namely the apparent misnumbering of months like December. Everyone familiar with the decimal system knows that *deci–* is a prefix meaning "ten" and that December should be the tenth month. But in fact, it is the twelfth month. Why then isn't it named something like "Duodecember"?

March, as the name of a month, is equally curious when one begins to think about it, but for different reasons. Not only is it the third month of our calendar, but as a noun it may be as lively as John Philip Sousa's "Washington Post March," which popularized the two-step, or as impressive as Schubert's "Marche Militaire" in D-major for four hands! As a verb, to march means "to advance or move forward." The proverbial "march of time" is but one expression showing how these ideas may come together.

What should one make of these kinds of disparate connections? Are they at all related? As it turns out, there is an explanation that accounts for both March and December. Moreover, the answer in part is mathematical, for it depends, in a very basic way, upon how we have come to count or number the months of the year.

Keeping time, as T.S. Eliot has written, is not only a musical expression, a matter of keeping the rhythm, but it is at its most fundamental and historical, a matter of seasons, of constellations. Not surprisingly, ancient civilizations kept track of time with the most obvious means at their disposal—the predictable change of seasons, the cycles of the moon, the daily, and annual motions of the sun.

Significant times of the year were often associated with the appearance at sunrise of certain stars; the heliacal appearance of Sirius in the summer sky, for example, marked the beginning of the "inundation" season for the Egyptians when the Nile would overflow its banks, annually rejuvenating the land, a crucial agricultural event each year.

Similarly, the easily observed phenomena of the solstices and equinoxes have played major roles in the religious observations of numerous cultures, with the vernal equinox associated with spring and the reawakening of new life being the most important. Thus, many cultures begin the new year understandably enough with the advent of spring.

Astronomy and mathematics have played complementary roles in determining how we go about accounting for this sort of time, and the advance of the seasons, in a variety of different ways. The ancient Romans, for example, counted their year quite literally by months, and vestiges of this method are preserved in our names for the last four—September, October, November, and December—all clearly related to the Latin seven, eight, nine, and ten. But these are all wrong by two, since September is the ninth month, and December is the twelfth. What accounts for this arithmetic discrepancy? Naturally enough, the answer goes back to the way in which the Romans first counted their months, and how they started the count to begin with.

The first eight months of our present calendar have no apparent numerical connections, but closer inspection reveals that there is more to January, March, and April in particular than meets the eye. Names often tell us a great deal about the objects they represent, and the early months of the year prove to be no exception. Apart from the obvious mathematical significance of months like September and December, there are more subtle mathematical connections with several other months of the year well.

To begin at the beginning, January is named after the Roman god Janus. This is especially appropriate, mathematically speaking, for a rather subtle reason. In Latin, the word *janus* means "gate" or "arched passage way," both suggestive symbols for beginnings. Similarly, the mythological figure of Janus was called the "doorkeeper of heaven" (thus we have an appropriate astronomical component for January). If time is really marked and measured in the heavens,

...Keeping time,
Keeping the rhythm in their
dancing
As in their living in the
living seasons
The time of the seasons and
the constellations....

—T.S. Eliot from
"East Coker" in *Four Quartets.*

what better place to begin the year than with January? Doorways when closed may signify the end, when open a new beginning.

Not only was Janus recognized as the god of beginnings and endings, but his blessing was asked at the start of every day, month, and year, also signifying firsts. Our word "year," in fact, is derived from an Indonesian-European root *ya* related to *janus*. Iconographically, Janus is often represented with two faces, one looking back to the old year just past, the other ahead, appropriately, as January begins the new year.

The second month, February, is less interesting from a mathematical point of view. It takes its name from the Latin *februa*, the Roman festival of purifcation, held on February 15 of each year. March was named for the god Mars (Martius in Latin)—more in a moment.

April is more interesting as we count through the calendar because it is misplaced. The month actually takes its name from the Latin *aprilis*, which is related to "latter" or "second." Thus, April should be the second month, not the fourth, just as December should be the tenth, not the twelfth, month. And indeed, although April is now the fourth, it was not always so. This is the first clue, as we number through the months, that the Romans did not always begin their count as we do now.

The oldest Roman calendar, in fact, once began with March, the month of Mars, best known as the god of war. Under a different attribute, however, the same god was also lord of agriculture, Mars Sylvanus, the god of spring vegetation. These may seem incompatible, if not contrary attributes. Nevertheless, there is a connection, arithmetically an ordinal one, for the earliest Roman calendar was determined agriculturally. Not only is this appropriate, it is to be expected in a largely agrarian society.

Naturally enough, the new year begins with a month associated with the growing season. Later, because of its proximity to spring and better weather, March also coincided with the opening season of military campaigns. Therefore, it is no mere coincidence that "march" may refer to a month, a military step or parade, and likewise to the stirring music by which armies march, whether in parade or off to war. Thus, both attributes—the agricultural and the martial—of the god Mars were appropriate to the first month beginning the earliest Roman calendar.

April, following March, was then in its proper sequence as the second month. Later, when the calendar was revised, with January and February added at the beginning, April retained its old name, as did the months from September to December. Their meanings, however, were thereafter quite inappropriate, and in fact, entirely wrong!

It is now clear why, today, December is misnamed. In the earliest Roman calendar, there were only ten months, beginning with March. Moreover, after April, May, and June, July was originally *Quintilis*, or the fifth month, and August was *Sextilis*, or the sixth month. September, October, November, and December then followed as they should have, in successive numerical order.

Only later, when Julius Caesar took to reforming the calendar, did he decide to rearrange things, renaming *Quintilis* as *Julius*, after himself (having been born in the month of Quintilis, he must have thought this entirely appropriate). Somewhat later, Caesar Augustus (not to be left out) changed *Sextilis* to *Augustus* in honor of corrections he authorized to the Julian calendar in 8 B.C. As a consequence of such reforms, when January and February were added at the beginning to make a twelve-month sequence, the names for April and the last four months of the older, ten-month year consequently fell out of their proper numerical sequence. Their names, however, were never changed to reflect the new order.

Equally puzzling as the order of the months and their names are the days of the week, their names, and the sequential order in which they appear, as well as the division of days into 60-minute hours and 60-second minutes. We shall examine their contributions to the measurement of time in the next issue of *Consortium*. ❑

Counting the Days... A Gift from the Gods

Solomon Grundy
born on a Monday

Christened on Tuesday

Married on Wednesday

Took ill on Thursday

Worse on Friday

Died on Saturday

Buried on Sunday

This was the end of
Solomon Grundy.

—Anonymous

JOSEPH W. DAUBEN

The events of Solomon Grundy's life followed a logical progression from birth to death. But is there anything logical about names like Wednesday or the order in which the days of the week appear? At first glance, the days of the week seem to exhibit no immediately discernable, rational order from Monday through Sunday. Nevertheless, their names actually reflect a coherent order. The sequence even conveys a hidden cosmological message, although it is one that is not immediately apparent.

If one knows where to begin, it is not difficult to understand why the days of the week follow in the order that they do. The reason, as will become clear in a moment, depends upon counting and cosmology, and dates back to the ancient Egyptians.

Like the months of the year, the days of the week also follow a repetitive pattern, a seven-day cycle; but as the days of the week progress from Sunday on through to Saturday, they do not reflect the sort of immediate mathematical pattern as September, October, November, and December do, with their nice progression of seven, eight, nine, ten.

There is, however, an obvious mathematical clue about the origin of the days of the week, and this is where we will begin our examination. At one time, the French calendar—adopted in the wake of the French Revolution and devised to sweep away all connections with the earlier, aristocratic, *ancien régime*—introduced revisions along purely rationalistic, nonsectarian lines, assuming a metric system to produce a ten-day week. Our week, however, has seven days, and the fact that the cycle repeats itself in periods of seven is significant.

Just as significant are the names given to each day of the week. While the meanings of Tuesday, Wednesday, Thursday, and Friday may not be apparent at first glance, Saturday, Sunday, and Monday are easier to define. The latter three clearly have in common the names of ancient deities (or the planets associated with the deities), namely Saturn, the Sun, and the Moon. Thus, we have Saturn's day, Sun's day, and Moon's day: Saturday, Sunday, and Monday.

The same is true of the other four days of the week, although the connections are not as direct in English. French and Spanish have not entirely obscured the original roots of Tuesday, which in French is *Mardi* and in Spanish, *Martes*. Both *Mardi* and *Martes* stem from Mars. In English, the word "Tuesday" is derived from the same idea, but here from Old High German. Tiu, according to Teutonic mythology, was the god of war and of the sky (equivalent to the god Mars), and lent his name to what in English has become Tuesday.

Wednesday, similarly, is *Mercredi* in French, or Mercury's day. The English version comes from the German god Wotan, and in Old Norse, Odin. Thus, Wotan's day eventually became Wednesday. So, too, the name Thursday arose. *Jeudi* in French is Jove's day, or Jupiter's day. English, however, follows a derivation again from the Old Norse god, Thor, or Thor's day. Finally, Friday is Venus's day, *Vendredi* in French. Our Friday comes to us from another female deity, the Old Norse goddess Frigg. Frigg's day has been transformed through its Middle English equivalent, *fridai*.

While the meanings of Tuesday, Wednesday, Thursday, and Friday may not be apparent at first glance, Saturday, Sunday, and Monday are easier to define.

Thus, the days of our week actually correspond to various gods and goddesses of the ancient religions and mythologies. Even so, is there any logical order to their sequence? Why progress from Sunday to Monday—Sun to Moon? Or from Tuesday to Wednesday—Mercury to Jupiter? Why Friday, Venus, then Saturday, Saturn? This seems a hopeless jumble, with no rhyme or reason.

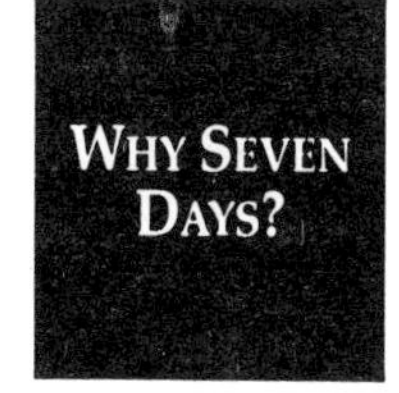

The fact that there are exactly seven days in the week, and not six, eight, or ten—or some other arbitrary number—provides a clue. The seven days are named, in fact, not just for any of the ancient deities, but for seven who were also identified with planets. In antiquity, exactly seven planets were known, seven moving objects in the heavens. The word "planet" means, in ancient Greek, "wanderer." To the ancients, the sun was a wanderer and hence, a planet, along with the others, Mercury to Saturn.

The ancient cosmology was also *geocentric*—Earth was thought to be stationary, virtually at the center of the universe. From this point of view, Earth was not a planet. The earliest of such detailed cosmologies was devised by the Egyptians, who regarded Earth as being at the center of the universe. The Greeks, and later the Romans, adopted this view and listed the planets as Mercury, Venus, Sun, Mars, Jupiter, and Saturn.

This followed in increasing order the relative distances of the planets from Earth. Basically, these distances were estimated by the time it took each planet to complete one circuit around the zodiac. Mars required a little less than two years, while Jupiter needed almost twelve years and Saturn more than twenty-nine years to complete a single circuit of the heavens and return to its starting point.

Egyptian Table of Days

		DAY ① SATURDAY	DAY ② SUNDAY	DAY ③ MONDAY	DAY ④ TUESDAY	DAY ⑤ WEDNESDAY	DAY ⑥ THURSDAY	DAY ⑦ FRIDAY
	1	Saturn ♄	Sun ☉	Moon ☽	Mars ♂	Mercury ☿	Jupiter ♃	Venus ♀
	2	Jupiter ♃	Venus ♀	Saturn ♄	Sun ☉	Moon ☽	Mars ♂	Mercury ☿
	3	Mars ♂	Mercury ☿	Jupiter ♃	Venus ♀	Saturn ♄	Sun ☉	Moon ☽
	4	Sun ☉	Moon ☽	Mars ♂	Mercury ☿	Jupiter ♃	Venus ♀	Saturn ♄
	5	Venus ♀	Saturn ♄	Sun ☉	Moon ☽	Mars ♂	Mercury ☿	Jupiter ♃
	6	Mercury ☿	Jupiter ♃	Venus ♀	Saturn ♄	Sun ☉	Moon ☽	Mars ♂
	7	Moon ☽	Mars ♂	Mercury ☿	Jupiter ♃	Venus ♀	Saturn ♄	Sun ☉
	8	Saturn ♄	Sun ☉	Moon ☽	Mars ♂	Mercury ☿	Jupiter ♃	Venus ♀
	9	Jupiter ♃	Venus ♀	Saturn ♄	Sun ☉	Moon ☽	Mars ♂	Mercury ☿
	10	Mars ♂	Mercury ☿	Jupiter ♃	Venus ♀	Saturn ♄	Sun ☉	Moon ☽
	11	Sun ☉	Moon ☽	Mars ♂	Mercury ☿	Jupiter ♃	Venus ♀	Saturn ♄
DAY	12	Venus ♀	Saturn ♄	Sun ☉	Moon ☽	Mars ♂	Mercury ☿	Jupiter ♃
NIGHT	13	Mercury ☿	Jupiter ♃	Venus ♀	Saturn ♄	Sun ☉	Moon ☽	Mars ♂
	14	Moon ☽	Mars ♂	Mercury ☿	Jupiter ♃	Venus ♀	Saturn ♄	Sun ☉
	15	Saturn ♄	Sun ☉	Moon ☽	Mars ♂	Mercury ☿	Jupiter ♃	Venus ♀
	16	Jupiter ♃	Venus ♀	Saturn ♄	Sun ☉	Moon ☽	Mars ♂	Mercury ☿
	17	Mars ♂	Mercury ☿	Jupiter ♃	Venus ♀	Saturn ♄	Sun ☉	Moon ☽
	18	Sun ☉	Moon ☽	Mars ♂	Mercury ☿	Jupiter ♃	Venus ♀	Saturn ♄
	19	Venus ♀	Saturn ♄	Sun ☉	Moon ☽	Mars ♂	Mercury ☿	Jupiter ♃
	20	Mercury ☿	Jupiter ♃	Venus ♀	Saturn ♄	Sun ☉	Moon ☽	Mars ♂
	21	Moon ☽	Mars ♂	Mercury ☿	Jupiter ♃	Venus ♀	Saturn ♄	Sun ☉
	22	Saturn ♄	Sun ☉	Moon ☽	Mars ♂	Mercury ☿	Jupiter ♃	Venus ♀
	23	Jupiter ♃	Venus ♀	Saturn ♄	Sun ☉	Moon ☽	Mars ♂	Mercury ☿
	24	Mars ♂	Mercury ☿	Jupiter ♃	Venus ♀	Saturn ♄	Sun ☉	Moon ☽

The Egyptians and Twenty-Four Hours

There is one last piece of information that we need before it is possible to explain the order of the days of the week, and this again brings us to the question of counting. Egyptians divided their days into hours, and how these hours were literally counted by the planets is significant.

Each day was divided into two periods, day and night. Each of these two major divisions was further subdivided into hours, twelve equal hours, thus giving a day a total of 24 hours. Each hour or "watch" of the day was assigned a planet. The first watch of the first day began with the planet farthest from Earth—namely, Saturn. The second watch was then assigned to the next closest, Jupiter, the third to Mars, the fourth to Sun, the fifth to Venus, the sixth to Mercury, the seventh to Moon.

1 Saturn
2 Jupiter
3 Mars
4 Sun
5 Venus
6 Mercury
7 Moon

The eighth watch of the first day began the cycle anew, with Saturn followed by Jupiter until the fourteenth watch ended with Moon.

8 Saturn
9 Jupiter
10 Mars
11 Sun
12 Venus
13 Mercury
14 Moon

The count then continued as before, until the twenty-fourth watch, belonging to Mars, ended the first day.

15 Saturn
16 Jupiter
17 Mars
18 Sun
19 Venus
20 Mercury
21 Moon
22 Saturn
23 Jupiter
24 Mars

Day one had completed its 24 hours, and it was now time to start day two. This, naturally, began with the watch following Mars, namely the Sun, and day two then continued its sequences of seven watches as before.

1 Sun
2 Venus
3 Mercury
4 Moon
5 Saturn
6 Jupiter
7 Mars
8 Sun
9 Venus
10 Mercury
11 Moon
12 Saturn
13 Jupiter
14 Mars
15 Sun
16 Venus
17 Mercury
18 Moon
19 Saturn
20 Jupiter
21 Mars
22 Sun
23 Venus
24 Mercury

So far, we have:
Day 1 beginning with Saturn;
Day 2 beginning with Sun;
Day 3 beginning with Moon.

Cycling through the planets in 24 hours, day three ends with Jupiter, and the fourth day's first watch is for Mars. Day five begins with Mercury, day six with Jupiter, and day seven with Venus. Day seven ends its twenty-fourth watch with Mercury, which begins day eight with the first watch for Saturn, and we have come full circle. Seven planets, seven days, and then a new week begins.

Thus, the Egyptians' cosmology and their division of the day into 24 watches or hours assigned to each of the planets account for what otherwise seems a jumbled sequence for naming the days of the week. Every time we refer to the days of the week, we are actually drawing upon ancient astronomical and mathematical conventions.

Another aspect of counting time also deserves to be mentioned here, because it too takes us back to an ancient source. Why do we divide time into hours of 60 minutes and minutes of 60 seconds? For this, we have another ancient civilization to thank—the Babylonians. They were a clever people who were also very mathematically minded. Systems of trade and exchange had grown up in Mesopotamia that had established units of 60 as a useful base for purposes of commerce and computation. Thus, there were 60 bushels in a *mana* and 60 *mana* in a *talent*, just as we have 10 cents in a dime and 10 dimes in a dollar. But 60 is a more satisfactory unit than 10 in one important respect—it is much easier to divide.

In fact, 10 can only be divided evenly by two numbers: by 2 and 5. If you want to divide 10 into thirds or quarters or sixths or twelfths, there will always be fractional remainders. But in day-to-day life, we often want to divide something into thirds, fourths, fifths, sixths—and it will help if the divisions come out evenly, with nothing left over. Sixty, because it is evenly divisible by so many numbers, including 2, 3, 4, 5, 6, 10, 12, and so on—gave the Babylonians considerable flexibility in their computations, whether applied for commercial reasons by accountants or businessmen, or for astronomical reasons by mathematical astronomers.

The Babylonians carried the same base 60 over to their measurement of time. Since astronomical calculations were done in base 60, each hour of the Babylonian day was divided into 60 minutes, and each minute into 60 seconds.

Consequently, in measuring time and in recording its passage, whether by months, days, minutes, or seconds, we are continuously indebted to the ancients who studied the motions of the planets, counted their cycles, and found that mathematics was a systematic key and useful device for helping to keep track of the passage of time. Moreover, the correspondence the ancients made between the planets and the days of the week underscores an important feature of counting that has always played a significant role in the history of mathematics, one that we shall explore in greater detail in the next issue of *Consortium*. ❑

From Here to Infinity

A Collection of Historical Notes on the Subject of Infinity

The Historical Notes column in *Consortium* featured the following three articles about infinity and its discoverer, Georg Cantor. We have combined them here into one series for easier use.

GEORG CANTOR
(1845–1918)

Georg Cantor was born in St. Petersburg, Russia, on March 3, 1845. When he was 11 years old, his family moved to Frankfurt, Germany, and by the age of 15, he was already demonstrating a talent for mathematics.

Cantor entered the University of Zurich in 1862, and after one year, he transferred to Berlin, where he majored in math and got his doctorate in 1867. As a faculty member of the University of Halle, he began his career as an unpaid lecturer, then was appointed assistant professor in 1872.

It was during this later time that he first began to tackle a problem on trigonometric series originally raised by Georg Riemann. Cantor's investigation of infinite sets of numbers led him to prove that the number of points within a square is no more "numerous" than the number of points on one of its sides. "I see it, but I don't believe it," he wrote in 1877.

In fact, many people didn't believe it, which led to personal, verbal attacks from other notable mathematicians. Although his career moved on to full professor in 1879, and his work in set theory had some appreciation, this criticism caused Cantor to suffer many nervous breakdowns and long bouts of depression.

Cantor died on January 6, 1918, in a mental hospital in Halle. Today, the power of set theory and its applications to broad areas of mathematics are fully accepted.

Infinity...
The Achilles' Heel of Mathematics?

Joseph W. Dauben

Two of the most fundamental concepts in mathematics—the twin concepts of counting and numbers—are also among the oldest. At first, progress was slow, and anything beyond one or two was simply regarded as "many." At this stage, notches carved in bones (there is a famous example of a wolf bone found in Vestonice, Czechoslovakia, bearing 55 notches) or slashes marked on the walls of Paleolithic caves (as at Alta Mira in Spain, Lescaux in France, or locations in Tanzania and other parts of southern Africa) sufficed for limited counting and rudimentary record keeping.

Incisions carved into Paleolithic bones like the one illustrated above, from the cave of La Mairie, Teyjat, Dordogne, France, show evidence of grouping and have been interpreted as tallying devices from which written number systems eventually developed.

Numbers doubtless arose simultaneously with speech itself, before any sort of written records were contemplated. Behind all such early attempts to count and number things was another essential idea of mathematics: the concept of one-to-one correspondence.

Counting on one's fingers is an easy example. Whether one counts a herd of cows or a bag of apples, if there is one cow or apple for each finger on both hands, there must be "ten." The result is also independent of the order in which the objects are counted. The last item counted is always the "tenth." In fact, given any *finite* set of *n* objects, counting them one by one in any order will always produce the same total number *n* of objects in the collection.

Counting, however, in its earliest phases was neither easy nor necessarily obvious. As in an earlier article, the ability to count was considered arcane in ancient Egypt, and the powers required to do so were considered supernatural and magical. Counting from one to ten was not even carried out directly, but facilitated through a mnemonic device by which each finger of the hand was associated with a number. Thus, numbers came to be associated by means of a one-to-one correspondence with the most natural sets of objects available for counting, primarily with the fingers of the human hand.

When the ancient arithmetician ran out of fingers, he went on to toes, bodies, and eventually had to generalize to facilitate counting very large numbers. Eventually, special symbols were devised to represent units of hundreds, thousands, and so on. This is all well and good for relatively small and finite numbers, but what happens if we think of trying to count collections that are not finite, but infinite in extent?

Consider, for example, the simplest case, the infinite set of all integers $N = \{1, 2, 3, ..., n, ...\}$. If we want to know "how many" integers there are, it is clear that the answer must be an "infinite" number; but to simply say that something is infinite is rather vague and therefore not very satisfactory, especially for mathematics.

Achilles and the Tortoise

The Greeks were the first to think seriously about the infinite, and they discovered very early that it was to be regarded as a tricky, usually contradictory concept. The famous paradoxes of Zeno of Elea (ca. 490 B.C.), which were closely related to the philosophical ideas of his mentor Parmenides, reflect some interesting yet troubling mathematical properties of the infinite. Best known, of course, is the paradox of Achilles and the tortoise. Assume that Achilles, swift of foot, is hoping to overtake the tortoise, who is some distance ahead of Achilles. The question now is how long will it take Achilles to catch up with the tortoise?

Zeno's answer is "never," and it doesn't matter how fast either Achilles or the tortoise is moving! To understand the logic behind Zeno's answer, consider **Figure 1**.

If Achilles at A_0 is behind the tortoise beginning at T_0, then in the time it takes Achilles to reach the tortoise's starting point at T_0, the tortoise will have advanced to position T_1, still ahead of Achilles. In the time it will now take Achilles to reach T_1, the tortoise will have advanced to T_2. In general, at any nth moment, Achilles will always be behind the tortoise.

Zeno viewed the problem of the infinite from another point of view as well. If an archer shoots an arrow toward a target, how long will it take the arrow to reach its goal?

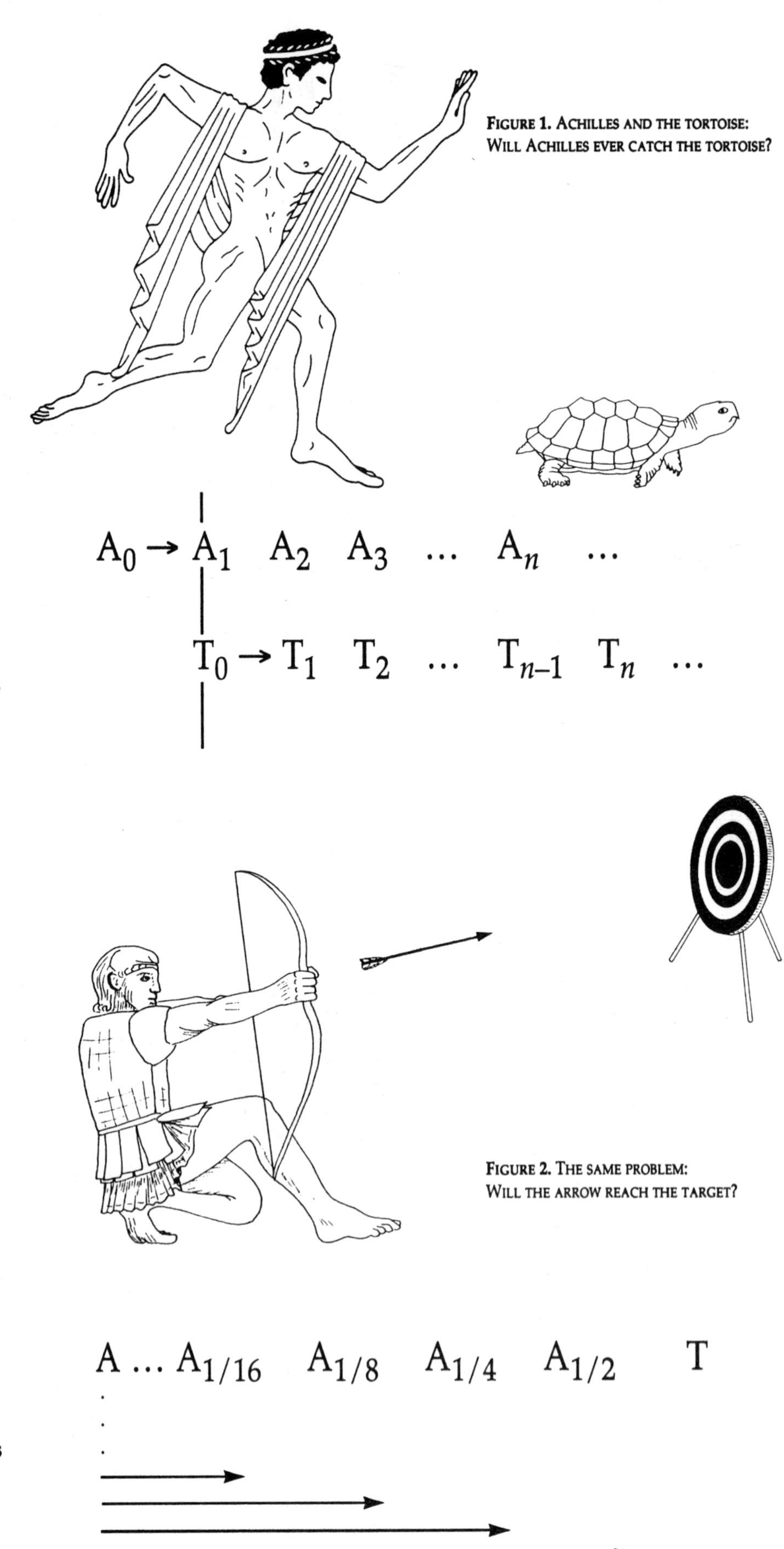

Figure 1. Achilles and the tortoise: Will Achilles ever catch the tortoise?

Figure 2. The same problem: Will the arrow reach the target?

Again, distance and speed are irrelevant to Zeno's answer, which is again, "never." This time, his argument takes on a somewhat different form. Consider, in **Figure 2**, the archer at A, the target at T, and the various distances the arrow must cover before it can reach the target.

Before the arrow can reach T, it must traverse half the distance AT, reaching $A_{1/2}$. But before it can reach $A_{1/2}$, it must traverse half the distance $AA_{1/2}$, reaching $A_{1/4}$. But before it can reach $A_{1/4}$, it must traverse half the distance $AA_{1/4}$, and so on. Since there are an infinite number of such successively halved distances that the arrow must traverse, it cannot even begin its trajectory from A toward T! Whatever Zeno's reasons may have been for devising these paradoxes, they served to warn mathematicians of the difficulties one might face in reasoning about the infinite.

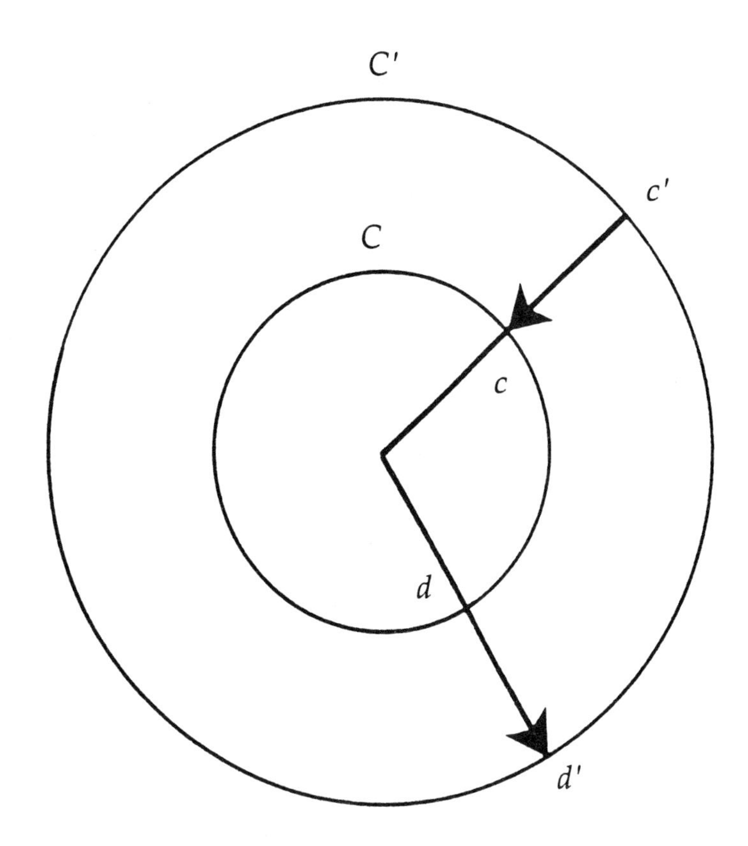

Figure 3. Two concentric circles.

More than a millennium later, at the beginning of the seventeenth century, Galileo also considered paradoxes of the infinite. Here, one example will suffice to show how the concept of one-to-one correspondence may provide a natural way to answer questions of "how many," even if the "many" involved are infinite.

In examining two concentric circles, Galileo knew that the respective lengths of their circumferences were different, depending on the relative radii of the two circles. However, if the circumferences of both circles are assumed to be composed of points, it is easy to see that both circumferences contain exactly the same number of points: a single radius drawn to the outermost circle directly associates exactly one point on each circumference.

Look at **Figure 3**. For any point c' on C', the radius drawn to c' will associate it with c on C. On the other hand, given any point d on C, the radius drawn through d and extended to C' will associate d with d' on C'. The seemingly paradoxical result is that while the length of the circumference $C' > C$, the number of points comprising C' and C must be the same!

It might now seem that the concept of one-to-one correspondence itself leads to some peculiar results, and that it might not prove to be a satisfactory means of answering questions like "how many" points there are on a given line. Somehow, it doesn't seem right that C' and C should have exactly the same number of points, despite the one-to-one correspondence between the two.

The first person to provide a viable solution from a mathematical point of view to these sorts of difficulties concerning the infinite was a German mathematician, Georg Cantor, who was born in St. Petersburg in 1845. He was educated, however, at the University of Berlin and spent the rest of his life in Germany teaching at the university in Halle. Cantor, as we shall see, took the idea of one-to-one correspondence, which was historically natural in developing the idea of counting, and applied it to infinite sets in a way that proved successful, even though it met with some adamant opposition at first. ❑

Counting to Infinity & Beyond
Cantor's Discoveries

JOSEPH W. DAUBEN

How many positive integers are there? How many odd numbers? Are there more even numbers than there are numbers divisible by 17? And how, to begin with, can we make sense of such questions about infinite collections of numbers—or of anything that is composed of an infinite number of things—of fractions, points, line segments, etc.?

In the previous issue of *Consortium*, we explored several perplexing paradoxes of the infinite, which for centuries convinced mathematicians that the infinite was something best left out of mathematics. To include it seemed to invite trouble, and yet, without it, mathematicians seems to have no answers to many interesting and important questions.

The first person to explore the problem of infinity in mathematics in a successful and coherent fashion was the German mathematician Georg Cantor (1845–1918). He is known not only as the father of set theory, but as the creator of transfinite numbers, which for the first time made it possible for mathematicians to speak accurately about infinity and to distinguish between different orders of the infinite. For it turns out that some things are "more" infinite than others; to say that a set of objects is merely "infinite" is not being nearly specific enough. There are, in fact, different kinds of infinity. To start at the simplest level, consider the smallest infinite set, the set of all positive integers N. This set contains two distinct subsets, one of even numbers E and the other of odd numbers O. Both E and O are of the same size, for we can easily match one-to-one every even number with an odd number, using the idea of one-to-one correspondence to "count" simultaneously all members of each set:

$$\begin{array}{l} E = \{2, 4, 6, \ldots, 2n, \ldots\} \\ \quad\ \updownarrow\ \updownarrow\ \updownarrow \quad\quad \updownarrow \\ O = \{1, 3, 5, \ldots, 2n-1, \ldots\} \end{array}$$

Every even number, say 1562 in E, can be matched with exactly one odd number—in this case, with 1561 in O. Now combine both of these sets. Together they constitute the infinite set of all positive integers:

$$N = E \cup O = \{1, 2, 3, \ldots, n, \ldots\}.$$

It is now natural to ask a question: Is the set N of all the positive integers twice as large as the set of just the even (or odd) numbers alone?

At first glance, one might be tempted to say that the answer must be yes. Surely $N = E \cup O$ makes this clear: there must be twice as many integers (even and odd together) as there are even numbers alone. But it is important to remember what is meant by "number" when we ask "how many" objects there are in an infinite set. We mean, in fact, that two sets are of the same size if they can be placed in one-to-one correspondence with each other. And with a little ingenuity, it is not difficult to show that there are exactly as many even and odd numbers together as there are even numbers alone.

Consider the simple correspondence $n \leftrightarrow 2n$:

$$\begin{array}{l} N = \{1, 2, 3, \ldots, n, \ldots\} \\ \quad\ \updownarrow\ \updownarrow\ \updownarrow \quad\quad \updownarrow \\ E = \{2, 4, 6, \ldots, 2n, \ldots\}. \end{array}$$

For every even number $2n$, there is exactly one integer n to which it corresponds, and the two sets are actually the same "size." Cantor called all infinite sets that could be "counted" by placing them in one-to-one correspondence with the set of positive integers N *denumerable*. Among such denumerably infinite sets are the set of all odd numbers, and any set that is constructed through successive multiples of the integers. For every tenth number, or every number divisible by 17, for example, the following correspondences show that both sets are denumerably infinite:

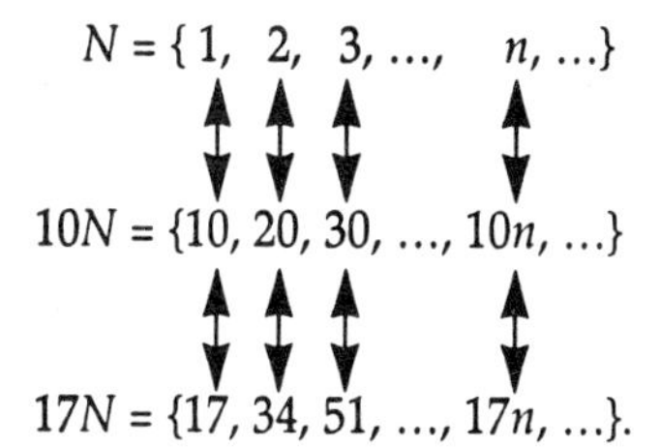

Consequently, the set of numbers {10, 20, 30, ...} is not one-tenth as large as the set of all integers, but is exactly the same size; both are denumerably infinite.

One of the first questions Cantor asked himself was whether or not all infinite sets were denumerably infinite. Could there be infinite sets larger than denumerably infinite ones? If so, what were these larger infinite sets? One likely candidate for such a "larger" set might be the set of all fractions Q. Try to picture them as represented by points on a line (as in **Figure 1**); it becomes immediately clear that there seem to be infinitely many more fractions than there are integers.

0 ¼ ½ ⅔ 1 2
1/10 ⅓ 7/12 ¾

FIGURE 1. FRACTIONS ON A LINE.

What makes the fractions seem so much more numerous than the integers alone is the fact that they are "dense"—

> Mathematicians were convinced for centuries that the infinite was something best left out of mathematics. To include it seemed to invite trouble, and yet, without it, mathematicians seem to have no answers to many interesting and important questions.

between any two fractions there is always another one (in fact, infinitely many more). Between 0 and 2, there is only one integer, 1. On the other hand, between the two fractions 1/2 and 3/4, for example, there is 7/12.
In general, between any two fractions a/b and c/d, there is always the fraction half way between them: $(1/2)((cb + ad)/bd)$, as well as infinitely many others, including $(1/n)((cb + ad)/bd)$, where n is any integer.

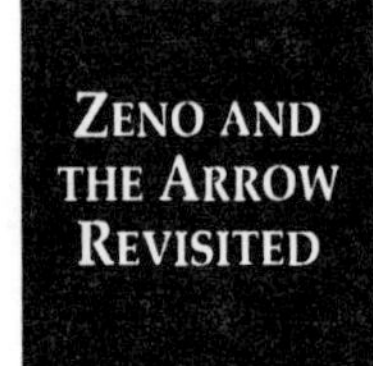

Zeno used a version of this property of the rational numbers, exploiting their denseness, to develop his famous paradox of the arrow (described on page 22), which led him to the conclusion that motion is impossible. If we now think of the arrow as being shot from point 0 toward a target at 1, then the arrow must first travel half the distance to 1/2, and half that distance to 1/4, and half that distance to 1/8, and half that distance to 1/16, etc. There is no end, or from Zeno's point of view, no beginning!

Not only are there an infinite number of such fractions $1/2n$, but between any two of them, there are infinitely many more fractions. This would seem to suggest that there must be infinitely many more fractions than integers, whatever we might mean by "infinitely many" more.

One way to show that the number of elements in Q is greater than the number of elements in N might be to show that there is no one-to-one correspondence between N and Q. The fact is, however, that Q is denumerably infinite, and that N and Q have the "same" number of elements.

To see this, consider the set of all fractions like 1/4, 2/3, 17/97, 34/59, or, in general, a/b (where a and b are any integers). By the nineteenth century, mathematicians had discovered the remarkable fact that the fractions Q were denumerably infinite, even though the set of all fractions in their natural order on the line was dense. Q was countable, meaning that the fractions could actually be placed into one-to-one correspondence with the integers. Consider, for example, the array in **Figure 2.**

denominators

numerators	/1	/2	/3	/4	...	/b	...
1	1/1	1/2	1/3	1/4	...	1/b	...
2	2/1	2/2	2/3	2/4	...	2/b	...
3	3/1	3/2	3/3	3/4	...	3/b	...
.							
.							
a	a/1	a/2	a/3	a/4	...	a/b	...
.							
.							

FIGURE 2. A SAMPLE ARRAY.

Any fraction at all, a/b, can be found in this array. It even contains all "duplicates," meaning fractions that would be equivalent if reduced to lowest terms, like $1/1 = 2/2 = 3/3 = \ldots = a/a$, or $2/3 = 4/6 = 8/12$, etc. The important feature of the array, however, is that it provides an immediate method for counting the fractions, duplicates and all. It is simply a matter of identifying each integer n of N with exactly one fraction a/b of Q. This is possible by simply winding through the array in a *boustrophedon*, or zig-zag fashion as in **Figure 3.**

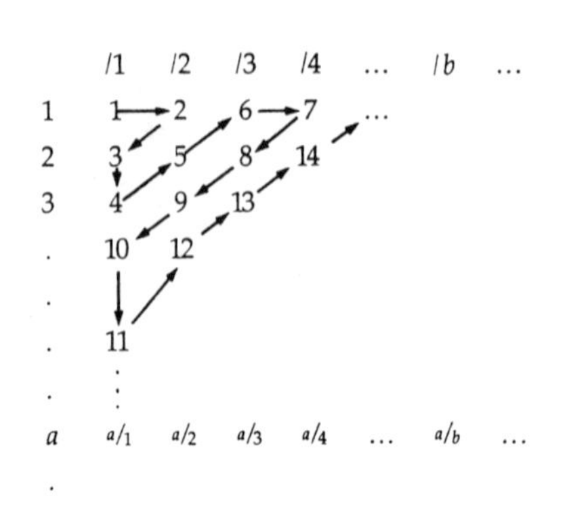

FIGURE 3. COUNTING THE FRACTIONS AND THEIR DUPLICATES.

Therefore, the set Q of all fractions is only denumerably infinite, and does not provide an example of an infinity of any greater magnitude than the integers themselves.

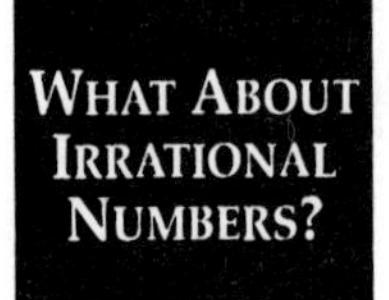

In search of a set absolutely larger than any of the denumerably infinite sets, Cantor, in an article published in 1874, turned his attention to yet another possibility. If the fractions were only denumerably infinite, what if one went on to the irrational numbers? Those like $\sqrt{2}$, Cantor knew, were among the "algebraic" numbers. A number is said to be algebraic if it is the root of any polynomial equation of finite degree and with integer coefficients. The set of all algebraic numbers A may therefore be regarded as composed of the solutions to all equations of the form:

$$a_0x^n + a_1x^{n-1} + \ldots + a_n = 0. \qquad \textbf{(1)}$$

Since any fraction a/b may be regarded as the solution of an algebraic equation, namely the linear equation $bx - a = 0$, the set Q is a denumerably infinite subset of the algebraic numbers. Irrational numbers like $\sqrt{2}$ or $\sqrt{5}$ are also algebraic, since they are solutions of the equations

$$x^2 - 2 = 0 \text{ and } x^2 - 5 = 0.$$

Although the details of Cantor's argument would require too much space for this column, it reduces to identifying every solution of an algebraic equation of the preceding form **(1)** with a specific number n of N, thereby establishing the size equivalence between N and A.

In search of a set
absolutely larger
than any of
the denumerably
infinite sets,
Cantor, in an article published
in 1874,
turned his attention
to yet another possibility.
If the fractions
were only
denumerably infinite,
what if one
went on to
the irrational numbers?

Having found that the sets of all rational and algebraic numbers Q and A were only denumerably infinite, what next? Are there no sets larger than these in an absolute sense? Cantor went on to investigate the set of all real numbers R, and in 1874, was able to publish, at last, his famous discovery that the real numbers were indeed nondenumerably infinite! His original method, one that is not so well known today, was both novel and remarkably simple. ❑

Infinity... and the Transfinite

JOSEPH W. DAUBEN

In a remarkable paper entitled, "On a property of the set of all real algebraic numbers" ["Über eine Eigenschaft des Inbegriffes aller reellen algebraischen Zahlen," *Crelle's Journal für die reine und angewandte Mathematik*, 77 (1874), 258–262] Georg Cantor showed that the set of all algebraic numbers[1] was only denumerably infinite. The method Cantor devised to show that the set of all algebraic numbers A was only countably infinite was a clever one, and relied upon the association of every algebraic equation with exactly one positive integer n, thereby establishing a one-to-one correspondence between the two sets $A \leftrightarrow N$.

[1] A number is said to be algebraic if it is the solution of a polynomial equation of the general form:

$$a_0x^n + a_1x^{n-1} + a_2x^{n-2} + \dots a_{n-1}x + a_n = 0.$$

Thus, $-1/2$ and $\sqrt{3}$ are algebraic numbers because they are the solutions of the algebraic equations $2x + 1 = 0$ and $x^2 - 3 = 0$, respectively.

Cantor was a pioneer in the study of the infinite from a rigorous mathematical point of view. In the early 1870s, he was especially interested in whether there might be infinite sets larger than denumerably infinite ones. Were there any infinite sets that could not be matched by a one-to-one correspondence with the set of integers N? Despite appearances, the infinite sets of rationals numbers and algebraic numbers were not larger than N, as Cantor was able to show.

Nevertheless, he persevered, and after the rational and algebraic numbers, he went on to try the set of all real numbers R. Here, at last, he was successful, and in 1873, he managed to show that the set R was indeed *not* denumerably infinite. This was a bold, even revolutionary, result.

An Oddity

It is rather odd, however, that Cantor did not broadcast this discovery in the title of his article on the subject, published in 1874. The result was there, but the title only mentioned his proof that the set of all *algebraic* numbers was denumerable. This, comparatively, was a conservative result. Only in passing, as it were, did he show at the end of his paper that the set of all real numbers R was non-denumerably infinite. This, he noted, also served to corroborate a proof Joseph Liouville had given in 1851, which proved there are an infinite number of *transcendental* numbers.

A real number is said to be transcendental of it cannot be expressed as the root of an algebraic equation; i.e., all real numbers that are not algebraic are called transcendental numbers. The two best-known transcendental numbers are π and e. We shall return later to Cantor's possible reasons for soft-selling his discovery about the real numbers, which in a short time would revolutionize mathematics. For the moment, let's concentrate upon *how* Cantor made his great discovery.

Cantor's Method

Most mathematicians are familiar with Cantor's famous method of diagonalization, by which he demonstrated that the set of real numbers is non-denumerably infinite. Cantor, in 1891, devised an ingenious method, which began by assuming that the set R *was* denumerable.

$$\begin{array}{llllll}
1 & 0.a_{1,1} & a_{1,2} & a_{1,3} \cdots & a_{1,n} \cdots \\
2 & 0.a_{2,1} & a_{2,2} & a_{2,3} \cdots & a_{2,n} \cdots \\
3 & 0.a_{3,1} & a_{3,2} & a_{3,3} \cdots & a_{3,n} \cdots \\
\cdot & & & & \\
\cdot & & & & \\
m & 0.a_{m,1} & a_{m,2} & a_{m,3} \cdots & a_{m,n} \cdots \\
\cdot & & & & \\
\cdot & & & &
\end{array}$$

FIGURE 1.

For example, consider the real numbers between 0 and 1 given as infinite decimals. If they were only denumerably infinite in number, then they could be listed via some one-to-one correspondence with the natural numbers N, as in **Figure 1**.

It is now possible to construct a number x that is *not* contained in this list. This can be done by going down the diagonal of numbers $a_{n,n}$, and then constructing x so that it will be different from any number m listed in the denumerably infinite array. Let $x = a_{x1}a_{x2}a_{x3} \ldots$ be constructed so that whatever $a_{n,n}$ may be, a_{xn} is 1 unless $a_{n,n}$ is 1, in which case let a_{xn} be 0. For example, suppose we have the following list of real numbers between 0 and 1 in **Figure 2**.

In order to construct x so that it cannot possibly occur in this denumerable list of real numbers, simply take $x = 0.0110\ldots10\ldots$. This x is a real number between 0 and 1, but it does not occur (it *cannot* occur in the list because of the way it has been constructed) as one of the real numbers in the correspondence with N. Consequently, R cannot be denumerably infinite since any denumerably listing of R fails to contain every real number.

This, however, was not the first of Cantor's proofs that the set of real numbers R is non-denumerable, but in essence was the second, which he devised in 1891. When he originally discovered the non-denumerability of R in 1873, his approach was somewhat different, although just as ingenious as the later, now better-known (and more easily grasped) method of diagonalization.

1	0.14159... n ...
2	0.33333...
3	0.66668...
4	0.97813...
.	
n	0.68798... 5 ...
$n+1$	0.94576... 01...
.	
.	

FIGURE 2.

It is rather odd that Cantor did not broadcast [his] discovery in the title of his article on the subject, published in 1874. The result was there, but the title only mentioned his proof that the set of all *algebraic* numbers was denumerable.

What were Cantor's possible reasons for soft-selling his discovery about the real numbers, which in a short time would revolutionize mathematics?

COUNTING R

Both proofs depend on obtaining a contradiction from the supposed denumerability of the set of all real numbers R. With this in mind, Cantor began with the assumption that R could be counted out in a denumerable order:

$$\omega_1, \omega_2, \omega_3, \ldots, \omega_n. \quad (1)$$

He then set out to prove that in any interval (a, b), there is always a real number η not contained in the above list. The argument proceeds as follows. Let a' and b' be the first two elements of **(1)** that are in the interval (a, b), with $a' < b'$. Then $a < a' < b' < b$. Now consider the interval (a', b'). Let a'' and b'' denote the next two elements of **(1)** that fall within (a', b'). If there is none, then any real number in the interval (a'', b'') will be a real number not listed in **(1)**. For example, $1/2(a'' + b'')$ would serve as such a number η not listed in **(1)**. Thus, the original assumption that R could be enumerated turns out to be wrong. Consequently, R must be *non*-denumerably infinite.

However, in general, there will always be more elements of **(1)** that will fall within the interval (a'', b''), so the process of constructing successive subintervals will continue. Suppose at the nth subinterval no further elements of **(1)** can be found that fall within (a^n, b^n). Then, as before, any real number in (a^n, b^n)—say $1/2(a^n + b^n)$—will do to provide us with a number not contained in the denumerably infinite list **(1)**.

"...for the
first time in history,
a mathematician
had succeeded
in determining
actual differences
between
various levels
of infinity."

By now, it is clear that the only interesting case is the one in which the process of nesting successive subintervals (a^n, b^n), each within its predecessor, does not terminate. What happens now? In this case, because of the way the subintervals have been constructed from the elements of (1), the endpoints of the intervals represent two infinite sequences: the one of left endpoints $a, a', a'', \ldots, a^n$, and the other of right endpoints $b, b', b'', \ldots, b^n$. Since both of these are bounded and increasing for the a's, decreasing for the b's, let their limits be denoted by a^∞ and b^∞, respectively.

Suppose now that $a^\infty \neq b^\infty$. Then the interval (a^∞, b^∞) will contain real numbers not listed in (1). As before, $1/2\,(a^\infty + b^\infty)$ will provide one example of an element of R not listed in (1). On the other hand, suppose $a^\infty = b^\infty$. Let η be this number, i.e., $\eta = a^\infty = b^\infty$. We can now show that η cannot be listed in (1). Assume that it were listed. Then $\eta = \omega_n$ for some index n of (1). But ω_n cannot be an element of (a^{n+1}, b^{n+1}), which is one of the subintervals of (a^n, b^n), although $\eta = a^\infty = b^\infty$ is supposed to be contained in all of the intervals (a^n, b^n), including (a^{n+1}, b^{n+1}).

This contradiction establishes the conclusion, namely that the set of all real numbers R, contrary to our initial assumption, must be non-denumerably infinite. Consequently, R cannot be put into any one-to-one correspondence with N.

This, of course, was a revolutionary discovery. It meant that, for the first time in history, a mathematician had succeeded in determining the actual differences between various levels of infinity. Infinity was no longer a vague concept meaning "more than any finite quantity." There were absolute differences between infinite sets, and Cantor had taken the first step in determining at least two of them: those infinite sets that were only denumerably infinite on the order of the set of natural numbers N, and those that were non-denumerably infinite on the order of the set of real numbers R.

At the time, however, Cantor could not extend this result in any precise way to reveal the limitless extent of the hierarchy of different infinities that his later development of transfinite set theory would produce. This research, which he carried out primarily in the 1880s and 1890s, would help to make even more precise the distinctions that could be drawn between different infinite sets. Cantor's paper of 1874, however, dealt with only two cases, and he made no attempt to emphasize his exciting new discovery about real numbers in the paper's title.

On the contrary, the title he actually gave the paper, "On a property of the set of all real algebraic numbers," was a calculated understatement of what he had achieved. Why did Cantor prefer to hide the fact that his article actually succeeded in demonstrating the non-denumerability of the real numbers R?

This question, and a discussion of Cantor's later development of transfinite set theory (along with his transfinite numbers) is the subject of the next article. ❑

Mathematics by Decree

RICHARD L. FRANCIS

No one would question the important place accorded various premises and definitions in mathematics. In the words of the layman, "this is the way it is." Hence, in a select geometric setting, only one parallel can be drawn to a line through a point not on that line. Moreover, there is no greater debate as it concerns, say, the definition of a polynomial, or of a parallelogram, or of a derivative.

Examples from elementary arithmetic likewise prove abundant by which one can stress the frequent encounter with statements rigidly set forth and without demonstration. Among other things, addition and multiplication are subjected, as if by decree, to the commutative and associative properties that govern such operations in the field structure.

Critical elements in a rigorous development involve decrees or agreements. They consist of those essential notions called definitions, undefined terms, and axioms, all within a logical framework (Aristotelian or otherwise). The remaining element, that of theorems (consisting of derived or proved statements), is, of course, the ultimate in developing a mathematical system. More precisely, the greater the number of theorems, the better the system.

However, there are other statements bordering on the fine line separating axioms from theorems, and these constitute a different kind of challenge. Such statements may not lend themselves to proof in a select context. Little hesitancy may thus be encountered concerning an axiomatic or decree-like disposition. All too often, it goes unnoticed or unmentioned that certain of these agreements can be considered as definite, consistent extensions of previously agreed-upon laws or rules. Hence, a contrast stands out between broad and comprehensive assumptions (the major axioms) and other specific agreements that are actually suggested or motivated by a projection of rules or formulas.

Some provable statements are often dealt with axiomatically, simply because of the level of understanding of the group being taught. Thus, such declarations as "the empty set is a subset of every set" or "the angle sum of a triangle is 180 degrees" may appear as basic or assumed propositions. This seems reasonable, although reliance upon unproved theorems may contain instructional risk.

Again, some mathematical results are made by assumption (axioms) or by common agreement (definitions). It is easy to dispose of other relations, then, in so simple a manner (even when more elaboration or statement motivation is in order). What, then, prompts such pronouncements as the following?

> x^{-n} equals $1/x^n$, **or**
> $x^{1/2}$ equals $\sqrt{x}$, **or**
> if x is not 0, then x^0 equals 1, **or**
> $(-a)(-b)$ equals ab, **or**
> $(a)(0)$ equals 0, **or**
> zero factorial equals 1

Let's analyze the last of these statements, which is typical of those mathematical results established by decree and with little or no elaboration or motivational comment.

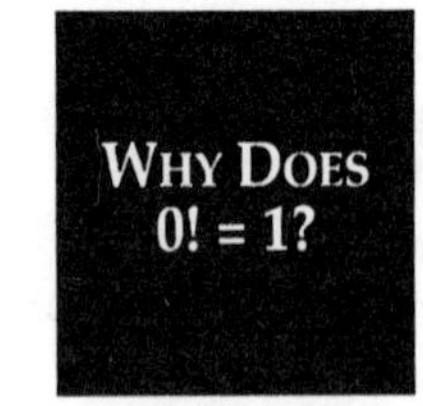

Accordingly, what leads the mathematician to equate (axiomatically or definingly) zero factorial to the number 1? Why not to 3 or 4, or even more tenably, to zero itself? If one can be so arbitrary and capricious as suggested here, mathematics may seem to make little sense to the student. Consider now the student's thought patterns, if n is a factor of n factorial (e.g., 3 is a factor of 3 factorial), then 0 is a factor of 0 factorial. Moreover, any expressed product with zero as a factor yields a zero result. Of course, the standard definition of n factorial (abbreviated $n!$) makes the case for n equal to zero meaningless. In this approach, n denotes the largest integer in

$$(1)\,(2)\,(3)\,(4)\,(5)\ldots(n).$$

How can n thus equal zero under these conditions? The definition further complicates the evaluation of 1 factorial, since multiplication—a binary operation—can scarcely be applied to a single element.

Under these conditions, is it any wonder that the teacher may want to take the easy way out, namely that of a decree? That is, zero factorial is 1, always 1, and never, never, must this emphatic declaration be questioned or probed as to why. And 1 factorial, by a similar logic, must also be 1.

Most significantly, the construction of larger and still larger factorials is determined by the formula

$$n! \text{ equals } (n-1)!\,(n).$$

Note that $5! = 4!\,(5)$ or $10! = 9!\,(10)$. Nothing remarkable stems from the formula if n is assigned integral values of 3 or more; however, if $n = 2$, the value of $1!$ (1 factorial) is suggested. That is, $2! = 1!\,(2)$ or $2 = 1!\,(2)$, or simply put, $1!$ is equal to 1.

Note that a broad and far-reaching formula is assumed applicable to these lesser n values. Though such a wide agreement is that of a mathematical decree itself, it is nevertheless of tremendous motivational value in the narrow setting of equating $1!$ to 1. The downward extension of a formula designed for n values greater than 2 has thus resulted in a consistent (and now to be expected) value for 1 factorial.

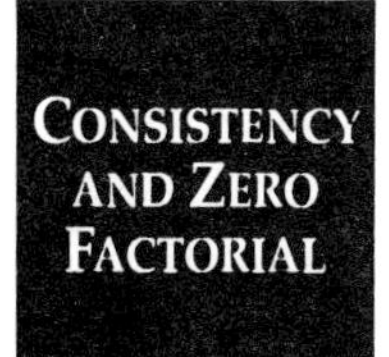

By an analogous procedure, can consistent meaning be given to zero factorial? Simply replace n by 1 in the extended formula

$$n! \text{ equals } (n-1)!\,(n),$$

so that $1! = 0!\,(1)$ or $1 = 0!\,(1)$ or $0! = 1$. This consistent extension of a formula proves intellectually satisfying to the student. It further suggests the reasonableness, the non-whimsicalness of equating zero factorial to 1. The dogmatic decree, however, that $0!$ is 1 leaves the capriciousness feature as a stumbling block to further inquiry (as in the binomial theorem, Pascal's triangle, combination and permutation formulas, or the evaluation of the number e).

A similar encounter with motivational development vs. the "dogmatic disposition of diverse decrees" (the "four-D" method) occurs with the exponent concept. The laws of exponents are initially presented with all exponent values restricted to positive integers. Only in time are these laws assumed to be binding on all exponent types. In such an extended setting, the rationale for assigning values to x^0 or x^{-3} or $x^{1/3}$ becomes apparent.

All of the decrees listed previously are positive in their wording (e.g., zero factorial always equals one). The negative counterpart also surfaces frequently. Hence, such negative decrees as "division by zero is impossible," or "the number 1 is not a prime" appear.

A few final words on the designation and the acceptance of key statements of relationships in mathematics are in order. Though the mathematician may seem to have unlimited freedom in the construction of definitions, the assignment of values, or the choice of axioms, there are various restraints that negate this. Admittedly, axioms today are viewed in a vastly different light from the "self-evident" approach of long ago. Still, perceptions based on experience often (but not always) guide the mathematician in this endeavor of choosing premises.

Though the mathematician may seem to have unlimited freedom in the construction of definitions, the assignment of values, or in the choice of axioms, there are various restraints that likely negate this.

Note, for instance, the assertion from Euclidean geometry that the straight line is the shortest path between two points. Moreover, premises and values must be selected in such a way as to avoid a contradictory system. One would hardly want to assume that a proposition is both true and false, or indirectly imply that 3 equals 4. Also, in the case of definitions, convention becomes important. Communication most assuredly breaks down when people try to convey ideas without any common agreement as to the meanings of words or symbols.

There are nevertheless those extended statements of a mathematical kind, some being fully and properly in the assumed category, that virtually speak out or demand some words of motivation. Whenever such a sharing of rationale is possible (acknowledging again the appropriate levels of understanding), it proves to be more than a powerful supplement; it becomes a vital part of an effective teaching/learning situation. Again, the unpleasant alternative is that of the rigidly worded, capriciously chosen, mathematical decree that may easily cause more problems than it will resolve. ❑

The Discoverer of Non-Euclidean Geometry: Lobachevsky, Bólyai, or Gauss?

Karen Doyle Watson

Who should be credited with the discovery of non-Euclidean geometry—Lobachevsky, Bólyai, or Gauss? Undoubtedly, each went to his grave considering himself to be the rightful owner of that honor.

The history of non-Euclidean geometry revolves around the great Carl Friedrich Gauss (1777–1855), who is considered to be one of the three greatest mathematicians of all time. Perhaps it was the perfectionist in Gauss (who refined his proofs and polished his manuscripts into flawless masterpieces) that prevented him from publishing his work on non-Euclidean geometry. Gauss's seal was a tree bearing seven fruits with the motto "few but ripe" [Faber 1983]. Or perhaps it was that the most highly revered and influential mathematician of the era chose to keep his work secret because of the unorthodoxy of his findings.

Although Kant encouraged the belief that Euclidean geometry was a necessary truth, Gauss realized it was impossible to demonstrate it. He feared the "clamour of the Boeotians" and refrained from publishing anything concerning the greatly discredited problem of parallel lines, yet he coined the term "non-Euclidean geometry" [Sommerville 1919]. The mystery enters when one realizes that three men, working independently and completely ignorant of each other's work, all had the same ideas.

Gauss and Wolfgang von Bólyai of Hungary had been friends since their student days at the University of Göttingen in Germany, and they discussed the subject of non-Euclidean geometry in 1798. Wolfgang's son Janos (1802–1860), a young army officer, wrote the following to his father about his work:

> *I have resolved to publish a work on the theory of parallels.... I have created a new universe from nothing. All that I have sent you 'til now is but a house of cards compared to a tower. I am fully persuaded that it will bring me honour....*
>
> [Carslaw 1916].

In 1832, Wolfgang von Bólyai published a small book entitled *Tantamen juventutum studiosam in elementa matheseos purae* (Essays for Studious Youths on the Elements of Mathematics), usually referred to as the *Tantamen* [Faber 1983]. At the back of the book, he placed an appendix concerning non-Euclidean geometry, which had been written by his son Janos. He had urged Janos to publish his findings promptly for the following two reasons:

> *First, because ideas pass easily from one to another, who can anticipate its publication; and secondly, ... many things have an epoch, in which they are found at the same time in several places, just as the violets appear on every side in the spring. Also, every scientific struggle is just a serious war, in which I cannot say when peace will arrive. Thus we ought to conquer when we are able, since the advantage is always to the first comer.*
>
> **[Bonola 1955]**

Janos Bólyai's opinion of the significance of his own work is expressed in an unpublished German version of the *Appendix*, in which he states that:

> *By the elucidation of this subject, one of the most important and brilliant contributions has been made to the real victory of knowledge, to the education of the intelligence, and consequently to the uplifting of the fortunes of men.*
>
> **[Carslaw 1916]**

No other published work of Janos Bólyai has ever been found.

Gauss was surprised when he received from Wolfgang a copy of the *Tantamen* with Janos's *Appendix*. Gauss described the younger Bólyai's work as elegant and concise. Gauss says that in the *Appendix*, he found "all my own ideas and results," and he judged the "young geometer vs. Bólyai as a genius of the highest order." However, Gauss wrote to the elder Bólyai, saying that he was "unable to praise this work" because Janos's results "have occupied my mind partly for the last thirty or thirty-five years." Hence, to praise Janos's work "would be to praise myself" [Carslaw 1916].

The elder Bólyai forwarded Gauss's letter to his son, stating that "Gauss's answer with regard to your work is satisfactory and redounds to the honour of our country and nation." Janos Bólyai's response, however, was one of chagrin and doubt, wondering whether Wolfgang had sent Gauss his early papers concerning non-Euclidean geometry. Janos held that Gauss had treated him badly; however, he later decided that his original suspicions were unfounded.

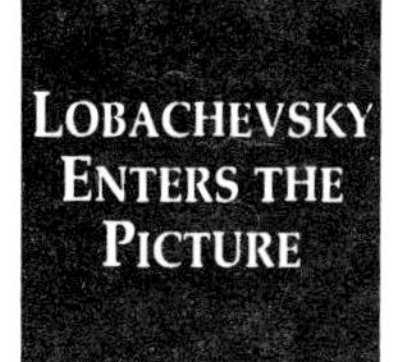

If Gauss's assertion is true, perhaps he should be acknowledged as the true discoverer of non-Euclidean geometry. But the coincidence of the proximity in the history of mathematics of the studies of Bólyai and Gauss is further confounded by the work of Nikolai Ivanovich Lobachevsky (1792–1856), a professor of mathematics at the far away University of Kazan in Russia. Lobachevsky was a pupil of Gauss's friend and countryman Bartels.

Unlike Bólyai, Lobachevsky was a prolific writer, whose first published work was the 70-page memoir in Russian entitled *On the Principles of Geometry* [Carslaw 1916]. As early as 1841, Gauss was aware of Lobachevsky's writings. He informed the elder Bólyai of the Russian's work in 1848, and the father passed the information along to his son. Janos was astonished by the close resemblance of Lobachevsky's *Geometrishe Untersuchungen zur Theorie der Parallelinien*, published in 1840 (in German), to his *Appendix*, which appeared in 1832 in Maros-Vasarkely.

Although Gauss was surprised by the remarkable agreement of the Hungarian and the Russian mathematicians' work, Janos Bólyai asserted that:

> "The nature of the absolute truth can only be the same in Maros-Vasarkely as in Kamschatka and on the Moon, or, in a word, anywhere in the world; and what one reasonable being discovers, that can also quite possibly be discovered by another...."
>
> [Carslaw 1916]

At the time of Lobachevsky's death in 1856 and Janos Bólyai's death four years later, neither had gained public recognition of his work, nor was either known by many mathematicians. There is no proof that Lobachevsky ever knew of Bólyai's work, though both men were directly or indirectly in communication with Gauss [Sommerville 1919].

Formal priority to the discovery of non-Euclidean geometry belongs to Lobachevsky, who was the first to publish his findings (in 1829), but Gauss was the first to acknowledge the idea of the existence of the new geometry (in 1816). Bólyai, however, seems to have appreciated the significance of the discovery more deeply than either Gauss or Lobachevsky [Yaglom 1979]. As Carslaw states in his history of the discovery of non-Euclidean geometry [1916]:

> *Had Gauss only made public reference to their discoveries, instead of confining himself to praise of their work—cordial and enthusiastic though it was—in conversation and correspondence, the world would earlier have granted them the laurels they deserved.* ❑

References

Bonola, Robert. 1955. *Non-Euclidean Geometry: A Critical and Historical Study of Its Developments*. New York: Dover Publications, Inc.

Carslaw, H.D. 1916. *The Elements of Non-Euclidean Plane Geometry and Trigonometry*. London: Longmans, Green, and Co.

Faber, Richard L. 1983. *Foundations of Euclidean and Non-Euclidean Geometry*. New York: Marcel Dekker, Inc.

Sommerville, D.M.Y. 1919. *The Elements of Non-Euclidean Geometry*. Chicago: The Open Court Publishing Co.

Yaglom, I.M. 1979. *A Simple Non-Euclidean Geometry and Its Physical Basis*. New York: Springer-Verlag.

What's In a Title?

Problems With Non-Denumerable Infinities

JOSEPH W. DAUBEN

In 1873, Georg Cantor, the celebrated German mathematician who invented set theory, wrote a short, three-page paper that was published the following year. It was titled "On a property of the set of all real algebraic numbers" (*Über eine Eigenschraft des Inbegriffes aller reellen algebraischen Zahlen*) [*Crelle's Journal* 1874]. As noted in a past issue of *Consortium*, this extraordinary yet understated paper signalled the beginning of a new era in mathematics.

No one scanning its title, however, would guess that this paper disclosed Cantor's revolutionary discovery of the non-denumerability of the continuum of real numbers. Instead, the article bore a deliberately misleading title, suggesting that its major result was a theorem about algebraic numbers, thus failing even to hint at the more significant point that the paper actually contained. What could possibly have prompted Cantor to choose such a wholly inappropriate title for what now, in retrospect, strikes any modern reader as one of the most important papers in modern mathematics?

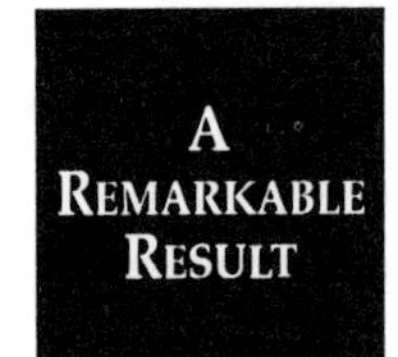

When Cantor originally wrote the paper late in 1873, he was nearly 30, but was only just beginning his mathematical career. Prior to joining the faculty of Halle University in 1869, Cantor had been a student at the University of Berlin. There he had studied with the greatest mathematicians of his day—Kummer, Weierstrass, and Kronecker. In Weierstrass's seminar, he had already used the method of one-to-one correspondences to show the equivalence of the set of rational numbers Q and the natural numbers N, establishing that Q is denumerable even though it is dense and the set of integers is not. This was the same device Cantor later used in paper in 1874 to show that the set of algebraic numbers is denumerable.

By 1872, Cantor was increasingly concerned with the structure of the continuum, largely because of the interesting questions raised by his uniqueness theorems (published between 1870–1872) for the representation of trigonometric series. Ultimately, Cantor succeeded in showing that the representations were unique even if an infinite set of exceptional points were permitted over the domain of definition.

This remarkable result depended upon the exceptional points being distributed in a very special way, so that they constituted what Cantor called *point sets of the first species*. (Given a point set P, Cantor defined the set of all of its limit points as P'. The set of limit points of P' constituted the second derived set P^2, etc. A set was said to be of the first species so long as P^n was empty for some finite value of n.) These were always denumerable sets. If he could show that the continuum was also denumerable (as the rational and algebraic numbers had proven to be), then he could characterize the continuum in terms of sets of the first species.
Although Cantor admitted in a letter to Dedekind (Dec. 2, 1873) that it might seem more reasonable to assume that no such one-to-one correspondence was possible between the real numbers R and the natural N, he could find no reason to make this assumption. On the other hand, he knew that if he could show that such a mapping were impossible, this would provide a new proof of Liouville's theorem for the existence of transcendental numbers.

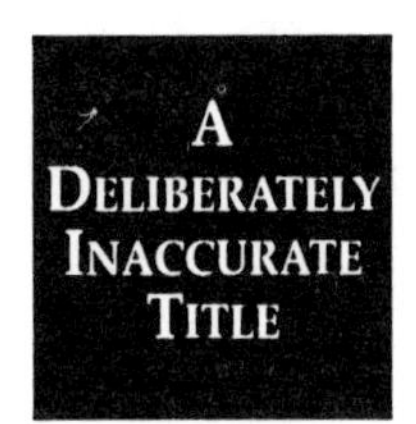

Within a few days, Cantor had finally found the answer, and wrote to Dedekind on December 7, 1873, explaining that it was not possible to provide any one-to-one correspondence between N and R. Cantor had a chance to show his proof to Weierstrass at the end of the month when he was in Berlin. Weierstrass was impressed, and urged Cantor to publish the result. Within weeks, the paper appeared in *Crelle's Journal*.

Which leads us back to the question of why Cantor would consciously have chosen a deliberately inaccurate title. One answer has recently been suggested by two German authors, Walter Purkert and Hans Joachim Ilgauds [Purkert and Ilgauds 1987]. They argue that Weierstrass was primarily interested in Cantor's result about the denumerability of the algebraic numbers; therefore, Cantor stressed this result in the title of his paper. Although Cantor told Dedekind of Weierstrass's interest in the denumerability of the rational numbers (a result Weierstrass soon applied to produce an example of a continuous, non-differential function), the real reason why he had given such a "restrained character" to his exposition was due, in part, he said, to conditions in Berlin. Whatever Cantor may have meant by that cryptic remark (which he never explained in further detail to Dedekind), what could it have had to do with the title of his paper?

The answer to this question hinges on another of Cantor's teachers at Berlin, Leopold Kronecker. Having studied with him, Cantor would have been well-acquainted with Kronecker's work in number theory and algebra, as well as with his philosophical views with respect to mathematics. By the early 1870s, Kronecker was already vocal in his opposition to the Bolzano-Weierstrass theorem, to upper and lower limits, and to irrational numbers in general. Kronecker's later pronouncements against analysis and set theory, as well as his adamant and well-known insistence upon using the integers to provide the only satisfactory foundation for mathematics, were simply extensions of these early views.

With this in mind, it is not unreasonable to suspect that Cantor had good reason to anticipate Kronecker's opposition to his proof of the non-denumerability of the real numbers. Certainly any result (such as Cantor's) that confirmed the existence of transcendental numbers—against which Kronecker's opinions were well-known—would not have pleased Kronecker.

Worse yet, Kronecker was on the editorial board of the journal to which Cantor submitted his paper. Had Cantor been more direct with a title, say, for instance, "The set of real numbers is non-denumerably infinite," or "A new and independent proof of the existence of transcendental numbers," he could have counted on a strongly negative reaction from Kronecker. After all, when Lindemann later established the transcendence of π in 1882, Kronecker asked what value the result could possibly have, since irrational numbers did not exist.

As Cantor contemplated publication of his article about the real numbers in 1874, an innocuous title was clearly a strategic choice. Reference only to algebraic numbers would have had a much better chance of passing Kronecker's eye unnoticed, for there was nothing there to excite either immediate interest or censure.

> No one scanning its title would guess this paper disclosed Cantor's revolutionary discovery of non-denumerability of the continuum of real numbers.

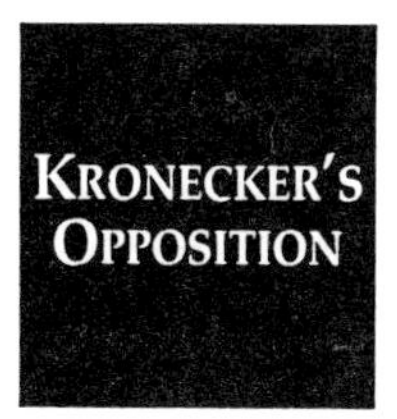

If the idea that Cantor may have harbored fears about Kronecker's opposition to his work seems unwarranted, it is worth noting that he had already tried to dissuade H.E. Heine, Cantor's colleague at Halle, from publishing an article on trigonometric series in *Crelle's Journal*. Although the article was eventually published, Kronecker was at least successful in delaying its appearance. (Heine was particularly vocal about this in letters to Schwarz, who was also a friend of Cantor's. Doubtless Schwarz and Heine would have both brought to Cantor's attention Kronecker's readiness—and ability—to block the publication of material with which he disagreed.)

Indeed several years later, Kronecker also delayed the publication of Cantor's next paper. This so angered Cantor that he never submitted anything to *Crelle's Journal* again. A decade later, he regarded Kronecker as both a private and a public menace—not only because he was condemning set theory openly, but also because he was condemning Weierstrassian analysis too.

There was a positive side, however, to Kronecker's opposition to Cantor's work, because it forced him to evaluate the foundations of set theory as he was in the process of creating it. This concern prompted long philosophical passages in Cantor's major publication of the 1880s on set theory, his *Grundlagen einer allgemeinen Mannigfaltigkeitslehre*, which was written in 1883. It was in this publication that Cantor issued one of his most famous pronouncements about mathematics, namely, that the essence of mathematics is its freedom. This was not simply an academic or philosophical message to his colleagues, for it carried as well a hidden and deeply personal subtext. It was, as he later admitted to David Hilbert, a plea for objectivity and openness among mathematicians. This, he said, was directly inspired by the oppression and authoritarian close-mindedness that he felt Kronecker represented, and worse, wielded in a flagrant and damaging way against those he opposed. Later, Cantor put his philosophy into action in a very concrete and lasting way. But that is another story and one we shall save for the next issue of *Consortium*. ❑

References

Cantor, Georg. 1874. Uber eine Eigenschaft des Inbegriffes aller reellen algebraischen Zahlen. *Crelle's Journal fur die reine und angewandte Mathematik*. 77: 258–262.

Dauben, Joseph. Counting to infinity ... and beyond. *Consortium* 21 (February 1987): 9–10.

Dauben, Joseph. Infinity and the transfinite. *Consortium* 22 (May 1987): 8–9.

Purkert, Walter, and Hans Joachim Ilgauds. 1987. *Georg Cantor*. Basel: Birkhauser.

"Let Freedom Ring!"

Georg Cantor and the Freedom of Math

Joseph W. Dauben

From almost the very beginning of his career as a professional mathematician, Georg Cantor had to face the opposition of Leopold Kronecker, who was at that time one of the world's greatest living mathematicians. Even before Cantor had begun to develop the full force of his theory of transfinite numbers, Kronecker's opposition was tangible. The effects of this opposition were two-fold. On one hand, it forced Cantor to present a philosophical defense of his new theory that was as carefully conceived as possible for a young mathematician—and for a theory that was only in its infancy at the time.

The tragic side of Kronecker's antagonism to Cantor's mathematics was largely personal: the extent to which it triggered and exacerbated Cantor's well-known nervous breakdowns. The first of these occurred in 1884, just as the tension between Cantor's new theory and Kronecker's opposition to it became critical (at least by Cantor's account).

It was at the same time that Cantor was beginning to experience his first technical difficulties with set theory. Not only was he unsuccessful in establishing the truth of his celebrated Continuum Hypothesis, but there is also evidence that early in the 1880s, Cantor may have just been becoming aware of the paradoxes of set theory, at least in terms of there being a largest ordinal number.

Paradoxes of Set Theory

For example, in his *Grundlagen* of 1883, Cantor referred to collections that are too large to be comprehended as a well-defined, complete, unified entity. Unfortunately, he wrote obscurely, with references to absolute sets couched in explicitly theological terms, explaining that "the true infinite or Absolute, which is God, permits no determination." In a footnote accompanying the *Grundlagen*, he explained that "The absolute infinite succession of numbers [*Zahlenfolge*] seems to me therefore to be an appropriate symbol of the absolute."

Was this a hint that he already understood the collection of all transfinite ordinal numbers to be inconsistent, and therefore not a set? Later, Cantor said that it was a veiled sign that, even then, he was aware of the paradoxical results that followed from trying to determine what transfinite ordinal number should correspond to the well-ordered set of all transfinite ordinal numbers.

In the 1890s, Cantor could no longer be so vague about absolute entities, and was forced to be much more explicit about the paradoxes resulting from the consideration of the sets of all transfinite ordinal or cardinal numbers. The solution Cantor then devised for dealing with such mathematical paradoxes was simply to bar them from set theory. Anything that was too large to be comprehended as a well-defined, unified, consistent set was declared inconsistent. These were "absolute" collections that lay beyond the possibility of mathematical determination. Cantor communicated all of this in letters first to Hilbert in 1897 and somewhat later to Dedekind in 1899. [Cantor's letters to Hilbert have been published by Purkert and Ilgauds in their book, *Georg Cantor*, Basel: Birkhauser, 1987, 224-227. The letters to Dedekind were first published by Zermelo in Cantor's *Gesammelte Abhandlungen* of 1932, reprinted by Springer-Verlag in 1982, 443-450.]

Whatever the extent of Cantor's awareness of the paradoxes may have been in the early 1880s, he was certainly sensitive to Kronecker's growing and increasingly vocal opposition. Above all else, it is clear that the philosophical concerns expressed in his *Grundlagen* were strategically crucial, in Cantor's opinion, for a comprehensive defense of his new theory. This was not only unusual at that time, but also is still unusual today. When Mittag-Leffler asked Cantor's permission to publish French translations of his papers on set theory in his newly founded journal *Acta Mathematica*, he persuaded Cantor to omit all of the philosophical portions of the *Grundlagen* as unnecessary (and possibly repugnant) to mathematicians who might find the theory of interest but the philosophy unacceptable.

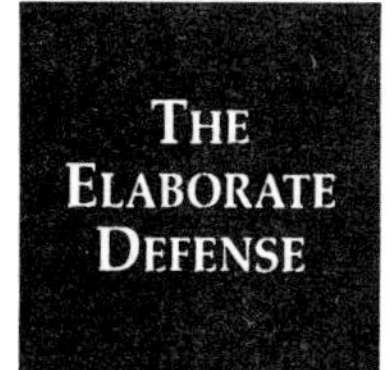

The philosophical arguments, however, were essential to Cantor, if not to Mittag-Leffler. They were essential because they were part of an elaborate defense he had begun to construct to subvert opposition from any quarter, especially from Kronecker. One ploy was to advance a justification based on the freedom of mathematics to admit any self-consistent theory. Applications might eventually determine which mathematical theories were useful, but for mathematicians, Cantor insisted that the only real question was consistency. This, of course, was just the interpretation he needed to challenge such an established mathematician as Kronecker. Cantor clearly felt obliged early in his career to plead as best he could for a fair hearing of his work. As long as it was self-consistent, it should be taken as mathematically legitimate, and the constructivist, finitist criticisms of Kronecker might be disregarded by most mathematicians for whom consistency alone should be the viable touchstone.

Cantor put his philosophy about the freedom of mathematics into action early in the 1880s, when his career had reached a point where he could do more than simply write about his feelings. During the 1880s, he began to lay the strategic foundations for an independent union of mathematicians in Germany. The specific goal of such a union, as he often made clear in his correspondence, was to provide an open forum, especially for young mathematicians. The union (as Cantor envisioned it) would guarantee that anyone could expect free and open discussion of mathematical results without prejudicial censure from members of the older establishment, whose conservatism might easily ruin the career of an aspiring mathematician. This was primarily needed in cases where the the ideas in question were at all new, revolutionary or controversial.

Cantor labored intensively for the creation of the *Deutsche Mathematiker-Vereinigung.* Eventually an agreement was reached and the Union of German Mathematicians held its first meeting in conjunction with the annual meeting of the *Gesellschaft Deutscher Naturforscher und Aerzte*, which met at Halle in 1891. Cantor was elected the union's first president, and at its inaugural meeting, he presented his now-famous proof that the real numbers are non-denumerable using his method of diagonalization.

The German union was not the end of Cantor's vision. He also recognized the need to promote international forums as well, and began lobbying shortly after the formation of the DMV for international congresses. These were eventually organized through the cooperative efforts of many mathematicians, and not directly, it should be added, as a result of Cantor's exclusive efforts. The first of these was held in Zurich in 1897; the second in Paris in 1900.

Promoting new avenues for discussion of mathematics was one way Cantor reacted to opposition of the sort his own research had provoked. Despite criticisms, especially from Kronecker, Cantor persevered, even in the face of his own repeated failure to resolve some of the most basic questions about set theory (notably his Continuum Hypothesis), and even though he had begun to suffer from increasingly serious cycles of manic-depression.

Ironically, as did his conflicts with Kronecker, Cantor's manic-depression may have served a useful purpose. In his own mind, his psychological state was closely linked to the infallible support set theory drew from his strongly held religious convictions. Letters (and the testimony of colleagues who knew him) reveal that Cantor believed he was chosen by God to bring the truths of set theory to a wider audience. He also regarded the successive waves of manic-depression that had begun to plague him in the 1880s—peaks of intense activity followed by increasingly prolonged intervals of introspection—as divinely inspired. Long periods of isolation in a hospital provided Cantor opportunities for uninterrupted reflection, during which he envisioned visits from a muse whose voice reassured him of the absolute truth of set theory, regardless of what others might say about it.

In promoting set theory among mathematicians, philosophers, and theologians (he even wrote to Pope Leo XIII at one point on the subject of the infinite), Cantor was convinced he would succeed in securing the recognition that transfinite set theory deserved. By stressing self-consistency and the intrinsic freedom of mathematics, he also advanced an essential element of any intellectual inquiry. The mind must be free to pursue the truth wherever it may lead. Inspiration should be encouraged, not confounded by arbitrary prejudice, and for Cantor, this meant a tolerance for theories judged upon standards of consistency and utility.

By now the utility of set theory has been established beyond question. So, too, the valuable lesson that no small part of its beauty, generality, and very great power lies in the freedom of the human mind to follow the implications of mathematical thought wherever they may lead. ❑

The Most Frequent Inauguration Day

RICHARD L. FRANCIS

Have you ever noticed that presidential inauguration day fell on Sunday in 1985? Do you recall perhaps that this also occurred at the beginning of President Eisenhower's second term of office in 1957? Can you possibly remember further back than this to the only other Sunday inauguration day of this century, namely that of Woodrow Wilson? It was Sunday, March 4, 1917, a troubled time just prior to our entry into World War I. As a matter of fact, only six times in the history of the United States has presidential inauguration day fallen on Sunday. What is the likelihood of Sunday inaugurals and when, if ever, will the two come together again?

LIKELIHOOD OF SUNDAY INAUGURALS

The coincidence of the two occurrences may seem somewhat of a rarity. However, such is not the case. Even as the thirteenth of the month falls more often on Friday than any other day of the week, so does January 20th in inauguration years fall more often on Sunday. Neither Friday the Thirteenth nor a Sunday inaugural is a calendar rarity, though such conclusions may come as somewhat of a surprise.

Mathematically speaking, the calendar is periodic. Its least cycle of perfect date-day repetition is that of 400 years. Christmas Day of the year 2389 (which is 400 years away) will fall on Monday, even as Christmas Day in 1989 falls on Monday. Accordingly, all that is required in answering the question of the frequency of Sunday inaugurals is to analyze a typical 400-year period of time. Basically, this reduces to simple counting. Other observations do assist, however, with the counting process. Within a given century, a shifting of five days forward occurs from one inaugural to the next. This is due to the four intervening years, exactly one of which is a leap year.

Inauguration Day of 1989, for example, is obtained by counting five days forward from Sunday, the day of the 1985 event. The result is Friday, as last occurred at the Kennedy inauguration of 1961. Caution must be exercised, however, in counting from one century to the next, as years ending in double zeros are not necessarily leap years.

Even as the thirteenth of the month falls more often on Friday than any other day of the week, so does January 20th in inauguration years fall more often on Sunday.

Inauguration Day, now fixed by the Constitution as January 20th in appropriate years, will occur on Sunday exactly sixteen times in the typical 400-year period, more than on any other day of the week. Tuesdays and Fridays follow closely, each occurring exactly fifteen times.

Sunday inaugurals are not so rare by the present pattern, but appear instead in some abundance. However, prior to the Roosevelt inauguration of 1937, the fixed inauguration date was that of March 4th. Quite remarkably, this date falls less often on Sundays and Tuesdays than on the remaining days of the week. That the founding fathers and mothers chose March 4th with the infrequent Sunday feature in mind is highly doubtful. Though the first presidential inauguration was scheduled for Wednesday, March 4, 1789, complications forced Washington's oath of office that year to be delayed until April 30th. In all other cases prior to 1937, presidential terms began on March 4th in inauguration years. Technically, inauguration years are those which number one more than a multiple of four.

Public ceremonies were held on Monday whenever March 4th fell on Sunday. Similar comments apply to the present scheme. Since January 20th fell on Sunday in 1957, President Eisenhower took the official oath of office that day in a private ceremony in the White House following church attendance. The public swearing-in occurred at noon on Monday, January 21, 1957. Likewise President Wilson took the oath of office privately in his official room in the Capitol on Sunday, March 4, 1917. The formal inauguration and the usual ceremonies took place the next day.

Though Sunday inaugurals prove rare by the March 4th scheme, they prove more abundant by the present, relatively young, January 20th arrangement. The long-standing March 4th plan makes the matter of Sunday inaugurals infrequent in the overall count.

The nineteenth century provided the first Sunday inaugural in the nation's history. This occurred with the ending of the first term of office of James Monroe. On Sunday, March 4, 1821, he succeeded himself to the high office. Zachary Taylor's only term (which he did not complete) began on Sunday, March 4, 1849 but the oath of office was taken on Monday, the 5th. Following a highly disputed election, Rutherford B. Hayes took the oath of office on the day following Sunday, March 4, 1877. As mentioned before, Woodrow Wilson's second term of office began officially on Sunday, March 4, 1917; Dwight D. Eisenhower's second term of office began officially on Sunday, January 20, 1957. These last two are the only instances of Sunday inaugurals in the twentieth century prior to 1985. All presidents taking the oath of office in Sunday inaugural years were incumbent except for Taylor and Hayes. The inauguration year of 1985 proved no exception.

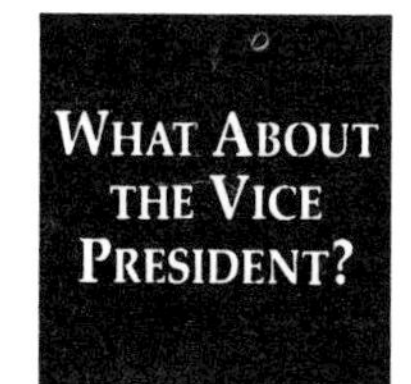

Though the primary focus is that of fixed inauguration days that occur on Sunday, it should be stated in passing that one (only one) of our vice-presidents succeeded on a Sunday to the high office following the death of a president. This occurred on Sunday, April 4, 1841 with the death of William Henry Harrison, and the automatic succession of John Tyler to the presidency. The oath was taken two days later. The remaining vice-presidential successions took place on other days of the week (Millard Fillmore on Tuesday, Andrew Johnson on Saturday, and Lyndon Johnson and Gerald Ford both on Fridays).

Inauguration Day now falls on Sunday more than any other day of the week. Yet, because of the long-standing March 4th tradition, Ronald Reagan became only the sixth president of the United States to have an inauguration day occur on Sunday. When will the event of a Sunday inaugural occur again? Not until the next century. The year will be 2013; it will be the beginning of the first or second presidential term of a man or woman now living. ❑

References

Francis, Richard L. 1985. Inaugural Daze. *Mathematics Teacher* 78: 66.

——. 1988. *A Mathematical Look at the Calendar*. HiMAP Module 10. Lexington, MA: COMAP, Inc.

The Notation For Inverse Trigonometric Functions: From Whence Did It Come?

KAILA KATZ

Calculus textbooks today usually introduce two notations, arcsin x and sin –1 x, for the angle whose sine is x, but generally choose to use the latter, an awkward notation that is difficult to justify. It is always accompanied by a statement such as: "the –1 is not to be regarded as an exponent, but rather as a means of denoting this inverse function." Kline, for one, has stated that the arc sin notation is really preferable to avoid ambiguity but takes more time and space to write [Kline 1967]. Questions that can be discussed to good pedagogical advantage are: how did the more ambiguous one ever emerge, who promoted it, and why?

The first designation for the inverse sine function was attributed to David Bernoulli in 1729, who used A.S. for the arc of the angle whose sine is x [Cajori 1952]. In 1769, the French mathematician Condorcet was using arc (sin. = x), and by 1772, Lagrange had shortened this notation to arc.sin. x.

Among those who considered such notation "long and inelegant" were Charles Babbage and John F.W. Herschel, both of whom became well-known in social and scientific circles throughout the British Isles, and to some extent on the Continent and in the United States. (Their translation of the widely-used Lacroix calculus text was adopted at West Point in 1842 [Simons 1931].) While students at Cambridge, they helped form a club there, the Analytical Society (1812–1813), to discuss and promote the superior works of the continental analysts. These works had been ignored in Britain since the Newton-Leibniz controversy over the priority of the ideas of the Calculus, to the detriment of the English-speaking academic communities.

The Nineteenth Century

Analysis in the 1800s referred to the branch of mathematics that used algebra and calculus rather than synthetic geometry to study curves and surfaces [Kline 1972]. The works of the eighteenth century continentals had led to a period of tremendous growth in analysis. By 1821, Cauchy, influenced by the works of his predecessors, gave the first satisfactory definitions of the derivative and the definite integral, definitions that had eluded mathematicians since the late 1600s. This work was virtually unknown in England, where the methods of synthetic geometry had remained dominant since the time of Newton out of loyalty to what were thought to be Newton's methods. The English who worked with mathematics had clung loyally to the Newtonian notation and the geometric approach to the calculus that Newton thought was necessary for the explanation of his ideas. Thus, the English generally remained aloof and isolated from the intellectual stimulation of the ideas that contributed to Cauchy's work.

Herschel had published an article in the *Philosophical Transactions* of the Royal Society of London in 1813 (at the age of 21) [Herschel 1813]. One whole page was a footnote explaining his choice of notation for the inverse trigonometric functions, such as $\cos.^{-1}e$, which he used in the body of the article. In that footnote, he makes use of the Leibnizian "d" notation for the derivative of a function. That symbol was not then in use in England, where the Newtonian term for the rate of change of the variable, *fluxion*, held sway and was notated by $\dot{x}$ (read: x dot). (Leibniz had chosen—after much deliberation, as was his custom—the "d," such as dy/dx, rather than the $f'(x)$ and y' used by Lagrange, the D of Arbogast, or Newton's dot [Cajori 1952].)

Herschel said in the footnote of his 1813 article that $\cos.^{-1}e$ must not be understood to signify $1/(\cos. e)$, but is meant to mean, what was written before as arc ($\cos = e$).

John F.W. Herschel
1792–1871

Son of William Herschel, the most eminent astronomer of the period, John Herschel graduated first in his class at Cambridge where he was a co-founder of the Analytical Society with Charles Babbage. John F.W. Herschel is best remembered for his work in developing analysis. He advocated the separation of the symbols of operations from the symbols of quantity, and operating on these symbols of operations as upon analytical symbols.

Charles Babbage
1791–1871

Charles Babbage was a 19th century English inventor and mathematician. He designed the Difference Engine to calculate and print multiplication tables. A forerunner of today's computers, it failed because parts could not be machined to the precision required. The computing element of the machine was a series of toothed wheels on a shaft that worked like a modern odometer. He pioneered scientific analysis of production systems.

He explained that if ϕ is the characteristic mark of an operation performed on any symbol x, then $\phi(x)$ represents the result of that operation. The notation $\phi^2(x)$ is the repetition of that same operation, i.e., $\phi(\phi(x))$. In the same way, d^2x is ddx and $\sin x$, $\log x$, etc., are merely characteristic marks of algebraic operations on x. Thus, $\sin^2 x$ should be $\sin \sin x$, and $\log^3 x$ should be $\log \log \log x$, etc.

Herschel then applied this reasoning to inverse functions: He says that if $\phi^n\phi^m(x) = \phi^{n+m}(x)$, and if $m = -n$, then $\phi^n\phi^{-n}(x) = \phi^0(x) = x$, that is, ϕ performed no times on x. Then $\phi^{-n}(x)$ must be such a quantity that its nth ϕ shall be x. In other words, $\phi^{-n}(x)$ must represent the inverse function applied n times.

At this point in the article, the form for expressing the function $\phi(x)$ is changed to ϕ, x to accommodate the extended functions, where Herschel let ψ be the characteristic symbol for the inverse function:

$$\phi^{-n}, x = \psi^n, x$$
$$\phi^n x = \phi^{2n}, \phi^{-n}, x = \phi^n (\phi^n, \psi^n, x).$$

Hence,

$$\psi^n, \phi^n, x = x.$$

Therefore,

$$\psi^{-n}, x = \psi^{-n}\psi^n\phi^n, x = \phi^{+n}, x..$$

For example,

$$d^{-1}V = \int V \text{ and } d^0V = V.$$

Similarly, said Herschel, $\sin^{-1}x$ is arc $(\sin = x)$, $\cos^{-2}x$ is $\text{arc}(\cos^2 = x)$, etc. (Leibniz used $d^{-1}x$ for $\int x$ in 1695, according to Cajori [1952]).

Herschel commented that it is easy to carry on this idea, and its application to many very difficult operations in the higher branches will "evince" that it is somewhat more than a mere arbitrary contraction. At that time, Herschel was doing work in the area of calculus of finite differences, and Babbage was working in the calculus of functions, which referred to the solution of general functional equations. Precise notation was vital to these topics, and both advocated the separation of the symbols of operations from the symbols of quantity, and operating on these symbols of operations as upon analytical symbols.

In Babbage's article "Notation," published in the *Edinburgh Encyclopaedia* in 1830, one gets a sense not only of the importance Herschel and Babbage attached to the role of notation in the development of an idea, but also of the development of analysis on the continent, the problems that faced the English mathematics community, and the generally futile attempts on the part of Babbage, Herschel and their friends to widen the mathematical horizons of their countrymen.

The essence of Herschel's footnote is also found in Babbage's article "Notation," which was published in the *Edinburgh Encyclopaedia* in 1830. This is a wide-ranging article, comprising six pages of double columns and small print, which covers ideas about notation that range from the trivial to the most sophisticated. His manner of exposition is quite dramatic, almost evangelical, and he refers to but does not explain the works of Euler and other continentals to which—as he well knew—his audience would have had little recourse. To read this article is to get a sense not only of the importance Herschel and Babbage attached to the role of notation in the development of an idea, but also of the development of analysis on the continent, the problems that faced the English mathematics community, and the generally futile attempts on the part of Babbage, Herschel and their friends to widen the mathematical horizons of their countrymen.

The efforts of these people were manifested in textbooks and other writings, and the study of mathematics did improve at Cambridge and in England by the latter part of the nineteenth century, but never to the level of the work of the Continentals. Herschel's footnote was one contribution to that effort. ❑

References

Babbage, Charles. 1830. Notation. *Edinburgh Encyclopaedia*. 18 vols. Edinburgh: W. Blackwood, 394–399.

Cajori, Florian. 1952. *A History of Mathematical Notations*. vol. II. Chicago, Illinois: The Open Court Publishing Company.

Herschel, J.F.W. 1813. On a Remarkable Application of Cote's Theorem. *Philosophical Transactions of the Royal Society of London* 103: 8–26, 10, footnote.

Kline, Morris. 1967. *Calculus: An Intuitive and Physical Approach, Part One*. New York: John Wiley and Sons, Inc.

——. 1972. *Mathematical Thought from Ancient to Modern Times*. New York: Oxford University Press.

Simons, Lao G. 1931. The Influence of French Mathematicians at the end of the Eighteenth Century Upon the Teaching of Mathematics in American Colleges. *Isis* XV: 1, 104–123, 120.

Like the Flight of the Bumblebee, Some Things are "Impossibilities"

RICHARD L. FRANCIS

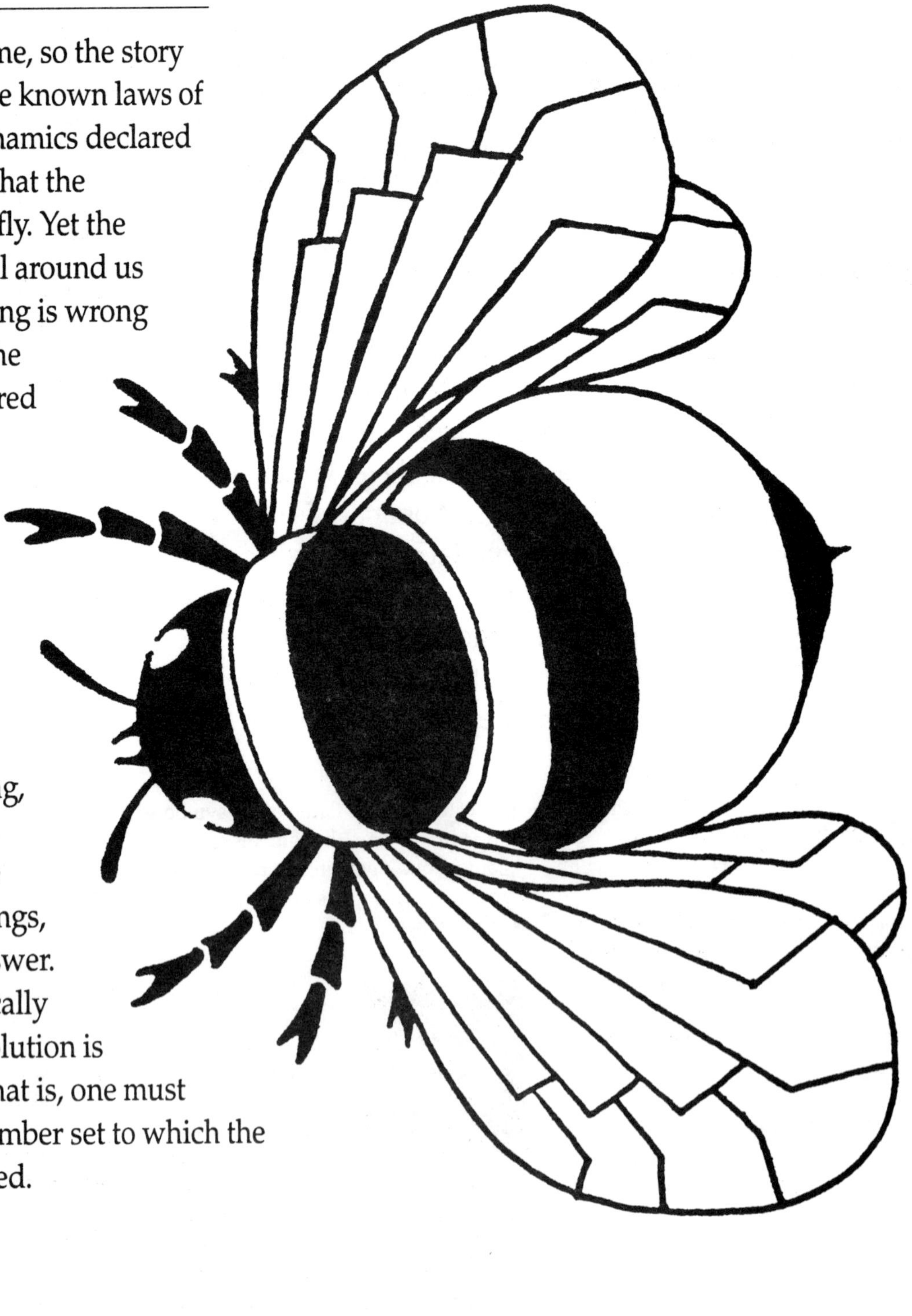

At one time, so the story goes, the known laws of aerodynamics declared boldly that the bumblebee couldn't fly. Yet the flying bumblebees all around us suggest that something is wrong somehow. Perhaps the student, too, has shared in such a conflicting dilemma. Surely all can remember that "two take away five" was declared impossible in the earlier years of mathematical training, formal or otherwise, but ultimately, in the more advanced settings, yielded a precise answer. To those mathematically well-versed, the resolution is extremely simple. That is, one must acknowledge the number set to which the discussion is restricted.

Even as the scientist may fail to account for the scope of the many variables affecting an outcome, so may the mathematician fail to mention the extent of the number sets surrounding his problem encounters. All too often, without regard to perspective, we find the student concluding "it can't be done." Such conflicts, stemming from operations being described simultaneously as possible and impossible, may well create the impression of mathematics as the science of contradiction. Arithmetic does not stand alone as the source of superficially confusing statements; algebra also must share in this distinction. As a matter of fact, virtually all of the mathematical disciplines contain quite an assortment of flightless bumblebees that flew anyway.

Mathematical concepts have arisen throughout history within very narrow domains of interpretation. The principle that "the concreteness of mathematics must precede the unfolding of the abstract" is well known. It explains the restricted settings invariably associated with mathematical beginnings. Even the child's formal classroom encounter with things mathematical parallels closely mankind's historical, motivational development of such.

Number class restrictions, generally rooted in concrete application, speak of the humble origins of various areas of exploration. The primary forms of number class extensions are today quite familiar. They proceed from the whole numbers and culminate ultimately in the set of complex numbers. This ultimate extension is motivated essentially by the objective of finding all solutions of algebraic equations.

Such a form of mathematics that ignores the conventional number domain restrictions will now be called an upheaval or cataclysmic form. Of course, the upheavals or cataclysmic forms often have a way of finding eventual acceptance, even as the earthquake, which drastically alters the terrain, is forgotten and the people routinely accept their new surroundings. Cataclysmic moments from the pages of mathematical history include such encounters as:

1. EXPONENTS.

The exponent concept initially fell in line with a counting approach. Thus, the expression x^3 indicated that x had been used three times in multiplication. Obviously, x could hardly be used –2 times or one-half of a time. Yet, a set of rules governing positive integral exponents was superimposed on other number types, so as to make them fully acceptable as exponents. Note the routine acceptance today of what had been revolutionary at one time, namely exponents of a vastly different kind.

2. LOGARITHMS.

It appears virtually axiomatic that negative numbers could not have logarithms. Seemingly, the number e or 10 raised to any power must result in a positive quantity. Still, due to the works of Euler and other forerunners in the study of complex numbers, it was discovered that $e^{\pi i} + 1 = 0$ (or, equivalently, $\ln(-1) = \pi i$). Such a shocking, nearly cataclysmic result is today a fundamental relationship in the study of functions of a complex variable.

3. FUNCTIONS.

The student establishes early the result that the sine of an angle can never exceed 1 nor be less than –1. Imagine the surprise that follows when the equation $\sin x = 2$ or the equation $\cos x = -3$ proves solvable. Of course, as before, compromising the number class restriction becomes necessary for such non-real solutions as these to make an appearance. Note the historical developments of complex analysis and the great or cataclysmic advancements necessary to admit the solvability of $\sin x = 2$.

4. FACTORIALS.

The factorial concept is relatively simple, provided one is calculating 3 factorial or 8 factorial. But how can meaning be given to zero factorial, or worse yet, to one-half factorial? Can base numbers in stipulating factorials be chosen from number sets other than the positive integers? The gamma function, attributed to Euler, Lagrange, and Gauss, generalized the factorial concept by using integrals and infinite products. As before, something of the cataclysmic form suggests itself here.

5. NUMBERS.

It was thought at one time that any number could be expressed as the root of an algebraic equation. In 1851, J.D. Liouville found a real number that could not be so expressed, thus giving rise to a cataclysmic number type called *transcendental*. Infinitely many such numbers are known to exist today, the most remarkable ones being π and e.

6. POWERS.

The student responds quickly to the rationale underlying the assertion that the number 9 has two square roots; that is, 3 times itself or –3 times itself will yield 9 for the answer. Such a rationale reveals just as quickly that –9 has no square root, at least not if the answer is chosen from the only number set (real numbers) thus far available. Nearly cataclysmic in history was the introduction into the number domains of mathematics of what is today called *imaginary numbers*. Needless to say, what at one time was regarded as unthinkable is today routinely accepted.

Perhaps other rock-ribbed statements of mathematics might deserve a second look in anticipation of a cataclysmic breakthrough. Though such is not always the outcome, the endeavor may well be worthy of attention.

Consider the last example above, and note that

$$\{x \mid x^2 < 0\} = \emptyset$$

allows for a far-reaching extension of number classes, that is, to the set of complex numbers. However,

$$\{x \mid |x| < 0\} = \emptyset$$

is always accepted as the final word on the subject of absolute values. That is, the absolute value of a number can never be negative. Such an emphatic statement appears to raise no further questions as to number class extensions. Why?

Evidently, the set of complex numbers fails to answer the question "what number has an absolute value of –1?" In other words,

$$|x + yi| = \sqrt{x^2 + y^2}$$

and proves necessarily positive or zero (never negative). Does there exist a new cataclysmic number type that might solve the equation $|x| = -1$? In pursuing the question, consider the introduction of an appropriate unit (a counterpart to $\sqrt{-1} = i$). Accordingly, let $|a| = -1$, where a is now to be called an *absolute unit*. Utilize as a premise the rule that the absolute value of a product is the product of the absolute values of the factors. Hence, in solving $|x| = -8$, write $x = 8a$ or $x = -8a$. That is, $|8a| = |8| \cdot |a| = 8(-1) = -8$. If the absolute value is positive, then solve as before. But what can be said about a? What kind of a number is it? Note that a is not real because the absolute value of a real number is non-negative. Note, too, that a is not even complex as the absolute value of a complex number is also non-

Arithmetic does not stand alone as the source of superficially confusing statements; algebra also must share in this distinction. As a matter of fact, virtually all of the mathematical disciplines contain quite an assortment of flightless bumblebees that flew anyway.

negative. Suppose a is a new number that leads to a far-reaching class Z of ultra-complex numbers.

Operations, both unary and binary, on the new number set could then be introduced, and questions pursued that relate to the number system. Such a procedure is quite reminiscent of the earlier development of the field of complex numbers. Perhaps too, a graphical approach could be taken, in imitation of the Gaussian plane for representing complex numbers, as well as some development of the concept of absolute values of ultra-complex numbers. For example, what would be the interpretation of $|x + ya|$? Even a counterpart to de Moivre's theorem for finding powers and roots might be pursued.

The story branches seemingly in all directions. It causes one to wonder what applications the new number concept might have in the world of mathematics.

Can new number forms be introduced just as simply and as easily as described here? One extremely important word of caution centers around consistency. If such an introduction gives rise to a contradictory system, the development must be rejected. Consider the problem of finding the absolute value of a^2.

Now $|a^2| = |a| \cdot |a| = (-1)(-1) = 1$. Then $a^2 = 1$ and $a = 1$ or -1, or $a^2 = -1$ and $a = i$ or $a = -i$.

But now $|a| = 1$ by this approach, certainly not -1 as the definition requires.

This apparent contradiction raises questions about the development based on an absolute unit a. The reader may wish to pursue this "system" more thoroughly, carefully defining the operations and the rules of the "game" in detail. Note that the logic that ruled out the absolute unit a above may have a counterpart in the imaginary unit i. Consider $i^2 = (\sqrt{-1})^2 = (\sqrt{-1})(\sqrt{-1}) = \sqrt{1} = 1$, which is erroneous, and $i^2 = -1$, which is correct.

The above reflects an effort to test the so-called "impossible." Other instances of delving into the emphatic realm of the "impossible" and in the quest of the cataclysmic include such encounters as in the "probability" definition. What kind of imaginary "event" could have a probability greater than one or less than zero? Could a highly theoretical probability be developed along the lines of such "outlandish" viewpoints, paralleling perhaps multi-valued logics that surpassed the dichotomies of Aristotle's logic? Or some other "unthinkable" modification such as a geometry of inordinately many (or fractional) dimensions, thus deviating considerably from the conventional "three"?

Of course, caution must be exercised whenever the mathematician uses the extreme declaration of "impossible." Somehow, there are associated with conclusions, meaningful conditions or qualifications that often remain unmentioned. It is to the careless in mathematics that these "bumblebees" of apparent conflict easily make an appearance. Some encounters are more subtle and challenging, however, whether the searcher is careful or not.

It was perhaps with a stinging sensation that the mathematician of yesterday busily realized that the impossible in the realm of numbers, may, by a consistent extension, result in a marvelous branch of study and application. No doubt, the challenge, in a myriad of forms, encircles us as flying bumblebees today. ❑

References

Richard L. Francis. 1986. From None to Infinity. *The College Mathematics Journal* 17:226–230.

——. 1988. Just How Impossible Is It? *Journal of Recreational Mathematics* 20: 4; 241–248.

Arithmetic in the Little Red Schoolhouse

JOHN P. POMMERSHEIM, PH.D.

Miss Firebrand rapped her pointer against her desk. The students reluctantly focused on the front of the room. Some of them experienced a sinking feeling of doom. "Please clear your desks," she ordered. "This morning we are going to have an arithmetic vocabulary test and I expect all of you to do well." She turned around and began writing on the board, using her most perfect script. A low murmur spread through the room as students began to realize the poor state of their preparation. One by one, the words to be defined appeared on the board, each one more difficult than the last. As she finished the list, Miss Firebrand slowly turned to face the class.

"You will have fifteen minutes to write a complete definition of each of these words on your slate," she said. "However, if any one of you is incorrect on more than half of the words, I will see you tomorrow morning at five o'clock to help me sweep the floors and stoke the stove." She glanced at the small watch that hung from her neck. "You may begin," she commanded. The students frantically began to write, rapidly glancing up and down from the board to their slates and back again to the board. Anguish was apparent on many faces, while others confidently wrote their definitions.

This vignette of days-gone-by may remind you of your own school days. But surely you can agree that the teaching and learning of arithmetic has undergone many changes over the past 100 years. Today children are exposed to fundamental concepts of arithmetic in early pre-school and beyond. In the nineteenth century, it was not unusual for children to stay in school for only four to six years. The learning of arithmetic consisted mostly of adding, subtracting, multiplying, and dividing whole numbers, possibly with additional work in fractions. After all, future farmers, carpenters, and homemakers needed no more arithmetic than that.

In those days, the methods of teaching arithmetic were very formal. Students were presented with rules and principles to follow, memorize, and apply to problems. Drill was widely used and oral teaching recommended. There was little understanding and enjoyment since the material was abstract and the student merely "learned" by doing.

Into the twentieth century, activity in the classroom remained focused on computation. The teacher usually demonstrated a procedure to the children and they in turn simply copied it until it was committed to memory. However, some of this drill was in the form of word problems, since computation was seen as a means to learning problem solving. Eventually the importance of students' understanding arithmetic, rather than just memorizing facts, was asserted. Drill was still considered important, but was used only after concepts were mastered through concrete and semi-concrete experiences.

The arithmetic textbooks used in the late nineteenth and early twentieth centuries provide interesting insights as to how a typical class was conducted. More often than not, the students did not have their own texts, especially in rural areas. The teachers had the only arithmetic book in class and followed it diligently. They tended to "teach the book" as it was written without elaboration. Meanwhile, the students did assignments on their slates or, in more affluent areas, in copybooks. Much student recitation of rules and principles took place in class. The students often learned to give the correct response to a question posed by the teacher, but they did not necessarily understand the response.

Since the teacher had only one text, it is evident that the reading level of such texts was much higher than what we find in today's arithmetic texts. For example, note the following instructions to the teacher indicating the steps (goals) to the learning of subtraction. These are taken from *The Elements of Arithmetic* by Edward Olney, published in 1877.

FIRST STEP: To ascertain the remainder when any number represented by one figure is taken from any number not less than itself, but less than itself plus 10.

To perform subtraction when minuend and subtrahend are represented by several figures each, and no figures in the subtrahend exceeds in value the figure in the same order in the minuend.

THIRD STEP: To perform subtraction when minuend and subtrahend are represented by several figures each, and there is a figure in the subtrahend whose value exceeds that of the corresponding figure in the minuend.

Another example is the following rule on how "to perform subtraction when both whole numbers and fractions are involved":

First take the fraction in the subtrahend from that in the minuend, or if the latter fraction be the smaller, from it increased by 1; and then subtract the remaining integers, prefixing the latter remainder to the former.

Similar examples abound. Although these examples come from the same arithmetic textbook, an investigation of other similar texts of that era reveal that the same level of reading was usually expected. Thus, vocabulary and the meaning of words was given high priority in arithmetic classrooms. After all, the teacher had to understand the information presented in the textbook, "translate" it, and present it to the students.

By the way, how well would you have done on Miss Firebrand's vocabulary test? Why not give it a try? Here is what she wrote on the board:

Give a complete definition of each of the following terms from our study of arithmetic:

1. separatrix
2. cipher
3. antecedent
4. consequent
5. involution
6. evolution
7. surd

I am confident that you will be successful in defining each of these words. If not, make sure your rooster is awake in time for you to be at the schoolhouse at five o'clock tomorrow morning! ❑

References

Morton, R.L. 1937. *Teaching Arithmetic in the Elementary School, Volume I, Primary Grades.* New York: Silver Burdett Co.

Morton, R.L. 1938. *Teaching Arithmetic in the Elementary School, Volume II, Intermediate Grades.* New York: Silver Burdett Co.

Morton, R.L. 1939. *Teaching Arithmetic in the Elementary School, Volume III, Upper Grades.* New York: Silver Burdett Co.

Olney, E. 1877. *The Elements of Arithmetic.* New York: Sheldon and Co.

Ray, J. 1905. *Ray's New Practical Arithmetic.* New York: American Book Co.

Smith, D.E. 1909. *The Teaching of Arithmetic.* Boston: Ginn and Co.

Stone, J.C. 1918. *The Teaching of Arithmetic.* Chicago: Benj. H. Sanborn and Co.

Suydam, M.N. 1982. Computation: Yesterday, Today, and Tomorrow. *Education in the 80's: Mathematics.* Washington, DC: National Education Association.

Answers to the vocabulary test:

1. separatrix: a decimal point.
2. cipher: another name for a zero.
3. antecedent: the first term of a ratio.
4. consequent: the second term of a ratio.
5. involution: the process of finding powers of numbers.
6. evolution: the process of finding roots of numbers.
7. surd: the root of a number that is not the product of equal factors (e.g., 2).

Quaternions:
The Key to Understanding Vectors

William Simpson

Vectors and the elements of vector analysis have, for so many years, been a standard topic within the undergraduate calculus program that it is easy to forget just how revolutionary the vectorial approach was. Calculus instructors can greatly enrich the understanding of their students by taking a few minutes to trace the development of vector analysis from the invention of quaternions by Hamilton to the controversial adaptations espoused by Gibbs and Heaviside. Such a short excursion would clear up such common questions as why there are two vector products (dot and cross), how anyone ever conceived such an arcane operation as the cross product, why the cross product cannot be directly generalized to four dimensional space, and whether there is a number system for number triples analogous to the complex numbers.

The story could begin with the Norwegian Caspar Wessel, who in 1799 first gave a geometrical interpretation of complex numbers of the form $a + bi$ by displaying them on the Argand plane as vectors extending from the origin to the point (a, b) and defining addition and multiplication in geometric terms. (It is interesting to note that Argand's independent but later (1806) treatment of complex numbers is usually cited as the first paper, whereas the protean Gauss probably beat both men to the discovery, having given strong evidence for this in a 1799 paper [Crowe 1985]).

Most of the mathematicians involved in the geometrical representation of complex numbers also endeavored to extend this approach into three-dimensional space. The fact that every such attempt met with failure made this problem one of some import.

By 1830, the mathematician William Hamilton had begun approaching this problem from an algebraic perspective. Considering complex numbers as number pairs (a, b), he algebraically defined addition as $(a, b) + (c, d) = (a + c, b + d)$, and multiplication as $(a, b) \times (c, d) = (ac - bd, ad + bc)$. Most importantly, he demonstrated that the geometric representation $a + b\sqrt{-1}$ and the algebraic representation (a, b) are equivalent and that if one can accept $(0, 1)$ as being a "number," then one in effect accepts the existence of $\sqrt{-1}$ since $i^2 = (0, 1)(0, 1) = (-1, 0) = -1$; this removed much of the mystery surrounding the notation i and gained a greater acceptance of the complex numbers [van der Waerden 1976]. In an 1873 paper, Hamilton noted that he was attempting to apply his algebraic approach to triplets $(a, b, c) = (a + bi + cj)$ to develop a three-dimensional number system.

Hamilton soon found that the greatest difficulty was one of finding a proper way to multiply number triples. Such multiplication had to be accomplished using term-by-term multiplication and at the same time preserving all the properties of the complex numbers, including the associative, communicative, and distributive laws as well as the law of the moduli ($|c_1 \cdot c_2| = |c_1| \cdot |c_2|$, length of the product vector equals the product of the vector lengths).

Off and on for a twelve-year period, Hamilton worked on this challenging problem. He convinced himself that $i \cdot j = -j \cdot i$ had to hold, but he could not seem to find any numerical value to assign to the product $i \cdot j$. Certain calculations suggested $i \cdot j = 0$, but this later led to a violation of one of the laws to be preserved. The conceptual breakthrough occurred on October 16, 1843. As he and his wife walked to a meeting of the Royal Irish Academy along the Royal Canal in Dublin, he suddenly realized that the product $i \cdot j$ could only work if it produced a third imaginary unit k, extending into the fourth dimension.

Once this idea had struck, he was quickly able to scribble down in the small notebook he always carried with him the various relationships connecting i, j, k and –1 [van der Waerden 1976]. As he later wrote to his son, "I [could not] resist the impulse—unphilosophical as it may have been—to cut with a knife on a stone on Brougham Bridge, as we passed it, the fundamental formula with the symbols, i, j, k; namely

$$i^2 = j^2 = k^2 = ijk = -1,$$

which contains the Solution of the Problem" [Crowe 1985]. He named his system of numbers the *quaternions*, after the squad of soldiers assigned to guard St. Peter, Acts 12:4 [Temple 1981].

The quaternions did not, however, represent the long-sought solution to the original problem. Not only did they require a four-dimensional space for their realization, but also, they violated one of the laws, that of communitivity, for $q_1 \cdot q_2 = -(q_2 \cdot q_1)$.

It was Hamilton's genius, strengthened by years of calculations, to intuit that one of the laws had to be abandoned and that the quaternions were probably the best system that one could devise. It was not until 1878 that Frobenious proved that no analogues of the complex numbers existed for any finite number of dimensions. Hamilton had indeed discovered the closest system to the complex numbers.

Van der Warten [1976] has suggested that Hamilton might have abandoned his research on quaternions if he had known about a simple counterexample published by Legendre, who demonstrated that the law of moduli does not hold in three-space by producing the example:

Sir William Rowan Hamilton
(1805–1865)

At the age of five, Hamilton could read Latin, Greek, and Hebrew and by nine, he had mastered another eight languages. When he was fifteen, Hamilton competed in public against a young American calculating genius and this experience shifted his interest to mathematics. In two years he had progressed through Newton's *Principia Mathematica* and begun to write mathematical papers. At eighteen, he entered Trinity College and achieved an unprecedented highest honors in both mathematics and English Verse.

In 1827, he was appointed both Professor of Astronomy at Trinity and Royal Astronomer at the Dunsink Observatory while still an undergraduate [Calinger 1982]. Hamilton's later accomplishments lived up to his early promise. His contributions to optics, dynamics, and mathematics made him one of the leading mathematicians of his day. Physicists remember him as the originator of a general dynamics principle embodied in the Hamilton operator but to mathematicians, Hamilton will always be linked to the quaternions.

$$
\begin{aligned}
\text{Let } (x, y, z) &= (1, 1, 1) \\
(a, b, c) &= (4, 2, 1) \\
3 &= 12 + 12 + 12 \text{ and} \\
21 &= 42 + 22 + 12;
\end{aligned}
$$

but $63 \neq u^2 + v^2 + w^2$, where u, v, w are bilinear in (x, y, z) and (a, b, c) with rational coefficients.

Luckily for future developments, Hamilton was unaware of this result. (In 1898, Hurwitz proved that in general, the Law of Moduli can only hold for dimensions of 1, 2, 4, and 8).

In working with his quaternions, Hamilton found that it was often convenient to view the number $a + bi + cj + dk$ as consisting of a "real" part a, and a "vector" part $bi + cj + dk$. It is at this point very illuminating to look more closely at the multiplication of two quaternions, denoted by ⊗.

$$
\begin{aligned}
q_1 \otimes q_2 &= (a_1 + b_1 i + c_1 j + d_1 k) \\
&\otimes (a_2 + b_2 i + c_2 j + d_2 k) \\
&= a_1 a_2 + a_1 b_2 i + \ldots \\
&+ (d_1 k)(d_2 k).
\end{aligned}
$$

Now, using $ij = k = -ij$, $jk = i = -kj$, $ki = j = -ik$ and rearranging terms we get:

$$
\begin{aligned}
q_1 \otimes q_2 &= a_1 a_2 - \overbrace{(b_1 b_2 + c_1 c_2 + d_1 d_2)}^{\text{dot product}} \\
&+ \overbrace{[a_1 (b_2, c_2, d_2) + a_2 (b_1, c_1, d_1)]}^{\text{multiplication by a scalar}} \\
&+ \overbrace{[(c_1 d_2 - c_2 d_1) i + (b_2 d_2 - b_1 d_2) j}^{\text{cross product}} \\
&+ (b_1 c_2 - b_2 c_1) k].
\end{aligned}
$$

If we write a quaternion as a combination of a real part r and a vector part v, we can view the multiplication more simply as:

$$
\begin{aligned}
q_1 \otimes q_2 &= (r_1 + \overline{v_1}) \otimes (r_2 + \overline{v_2}) \\
&= r_1 r_2 + [r_1 \overline{v_2} + r_2 \overline{v_1}] \\
&- (\overline{v_1} \cdot \overline{v_2}) + (\overline{v_1} \times \overline{v_2}).
\end{aligned}
$$

Thus, multiplication of scalars, multiplication of a vector by a scalar, dot product of two vectors, and cross product of two vectors are all components of a quaternion multiplication.

Feeling that the quaternions were one of the great discoveries in mathematics, Hamilton devoted the remaining twenty years of his life to developing his new system and demonstrating its usefulness.

By the time Hamilton died in 1865, quaternion methods were fairly well disseminated. Quaternions received a significant boost in 1873 when Clerk Maxwell used them in his *Treatise on Electricity and Magnetism*. Sensing the importance of the quaternions, Maxwell was very enthusiastic about their potential. He stated that quaternions provided mathematicians and scientists with a different and in many cases a superior "method of thinking."

Maxwell's enthusiasm spread to others but there occurred at this point a sharp difference of opinion as to the direction future development should take. Hamilton's followers urged the use of quaternions. Others, led independently by the efforts of the American physicist W. Gibbs and the self-educated but brilliant British engineer O. Heaviside, felt that the most important application of quaternion methods lay in three dimensions and so they sought a way to extract from the quaternions as much as would be made to work within three space [Crowe 1985].

By setting the real component of quaternions equal to zero, they worked with the vector component exclusively:

$$
\begin{aligned}
q_1 \otimes q_2 &= (0, a_1, b_1, c_1) \otimes (0, a_2, b_2, c_2) \\
&= (0 + \overline{v_1})(0 + \overline{v_2}) \\
&= (\overline{v_1} \times \overline{v_2}) - (\overline{v_1} \cdot \overline{v_2}).
\end{aligned}
$$

It was only a matter of time until the awkward quaternion notation $(0, a, b, c)$ was discarded in favor of the vector notation $(a, b, c) = ai + bj + ck$, and quaternion multiplication replaced by two separately defined operations of cross and dot product. These changes essentially created a new system—the vector space in three dimensions. From a purist standpoint, the vectorial system was deficient in many properties commonly considered to be important. Not only did the distributive and associative laws fail, but division was not defined. Furthermore, there was not one multiplication defined, but two. However, vectors lent themselves to applications in the physical sciences and engineering because they were defined in three-dimensional space and, most importantly, they had transformation and invariance properties that the quaternions lacked.

The usefulness of vectors was considerably enhanced when a German mathematician H. Grassman, in the process of developing the more general concept of tensor analysis, gave a geometrical interpretation to the cross and dot products by showing that

$$
\begin{aligned}
\overline{v_1} \cdot \overline{v_2} &= |v_1| \cdot |v_2| \cos \theta \\
\overline{v_1} \times \overline{v_2} &= \overline{\mu} |v_1| \cdot |v_2| \sin \theta.
\end{aligned}
$$

The debate between the proponents of quaternions and those of vectors raged for several decades. Gradually, however, practitioners in the scientific and engineering fields found that the preponderance of useful applications fell on the side of the vectorial system. By 1910, the elements of vector analysis, as it is taught today, were fully developed and accepted, while quaternions had begun a slide into relative obscurity.

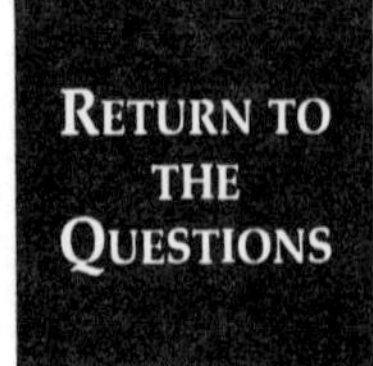

Returning now to the questions posed at the start of this historical summary, we find that they are all answered. The problem of developing a system of hypercomplex numbers of any degree higher than two is impossible; the nearest such system is the non-communicative quaternion system in four-space. The operation of quaternion

"Feeling that the quaternions were one of the great discoveries in mathematics, Hamilton devoted the [last] twenty years of his life to developing his new system and demonstrating its usefulness."

multiplication, which is a fairly straightforward generalization of complex number multiplication, can be viewed as a sum of terms, two of which are the familiar dot and cross product of vectors.

When vectors in three-space were created from the quaternions, it made good sense to define the cross and dot product separately and retain both of them because of their different geometrical interpretation and usefulness in solving applied problems. The origin of the cross product explains why it, by itself, cannot be extended to vectors of four dimensions; we must combine the dot and cross product and define the quaternion product in order to get a reasonable vector multiplication in four space.

The student who wishes to learn more of the interesting historical details surrounding the development of vectors should be encouraged to read Crowe's definitive work [1985]. ❑

References

Crowe, Michael J. 1985. *A History of Vector Analysis*. New York: Dover.

Temple, George F.J. 1981. *100 Years of Mathematics*. New York: Springer-Verlag.

van der Waerden, B.L. 1976. Hamilton's Discovery of Quaternions. *Mathematics Magazine* 49: 227–234.

The Nine Chapters on the Mathematical Art: An Amazing Book

FRANK SWETZ

Most secondary school students and teachers, if asked to name an important historical mathematical work, would probably respond: Euclid's *Elements*. The *Elements* were compiled by the Greek mathematician Euclid of Alexandria around 350 B.C. Euclid unified the various Greek mathematical discoveries of his time into a systematically organized work, employing definitions, axioms, and theorems. Since that time, the thirteen books that comprise the *Elements* have been extolled as a model of deductive reasoning. Geometry teaching up to the present day has been influenced by its contents. Indeed, the *Elements* have been a popular text—over a thousand editions of the work have been published since their first printing in Venice in 1482, earning the reputation of being the most well-studied books in the Western world other than the Bible. Certainly, Euclid's *Elements* has had a long and profound influence on Western mathematical thought and deserves to be acknowledged as a mathematical classic.

But there exists another ancient mathematics book, perhaps equally as profound in its mathematical influence and impact as the *Elements*, a book that remains almost unknown. It is the Chinese work, *Jiuzhang Suanshu*, the *Nine Chapters on the Mathematical Art*, a compendium of mathematical knowledge and techniques accumulated in ancient China up through the later Han Period (ca. A.D. 220). The obscurity of this book in the West rests on two factors: its remoteness to Western readers in general, as it has been translated into German and Russian but only exists in fragmentary English translations, a situation that is soon to be remedied; and lingering intellectual reluctance to recognize non-Western achievements. Despite this obscurity, even a brief survey of the *Nine Chapters* contents reveals that it is truly a mathematical treasure, one that tells us much about the uses and methods of early mathematics.

The extant version of the *Nine Chapters on the Mathematical Art* can be traced back to the scholar Liu Hui, who wrote a commentary on the work in approximately A.D. 263. He obviously made use of then existing versions of the book. From an analysis of the nomenclature and contents of the text, it appears that much of its material originated as far back as the Warring States period of Chinese history, 500 B.C. The book is a collection of 246 specific mathematical problems, compiled in question-and-answer form. Solution directions are rather terse and assume computations are performed using a set of computing rods and a counting board, traditional mathematical instruments. The problems, in turn, are arranged into nine chapters, each with its own mathematical emphasis and techniques:

CHAPTER 1

"Field Measurement" is devoted to land measure and the calculation of areas.

CHAPTER 2

"Cereals" discusses the use of proportions in the exchange of grain, i.e., rice, millet, etc.

CHAPTER 3

"Distributions by Proportion" continues a consideration of proportion in more difficult situations, some involving arithmetical and geometric progressions.

CHAPTER 4

"What Width?" involves area and volume computations and includes the extraction of square and cube roots.

CHAPTER 5

"Construction Consultations" is devoted to the determination of volume for a variety of geometric solids.

CHAPTER 6

"Fair Taxes" considers the collection of taxes based on population size and the distance from the capital. Problems of pursuit and allegation are also discussed.

CHAPTER 7

"Excess and Deficiency" presents the use of the "Rule of False Position" to solve simple linear equations.

CHAPTER 8

"Rectangular Arrays" develops matrix techniques for solving systems of equations. A problem with four equations and five unknowns is considered.

CHAPTER 9

"Right Angles" explores the use of the "Right Triangle Theorem," commonly called the "Pythagorean Theorem," to solve geometrically conceived situations. A quadratic equation is posed and solved in this chapter.

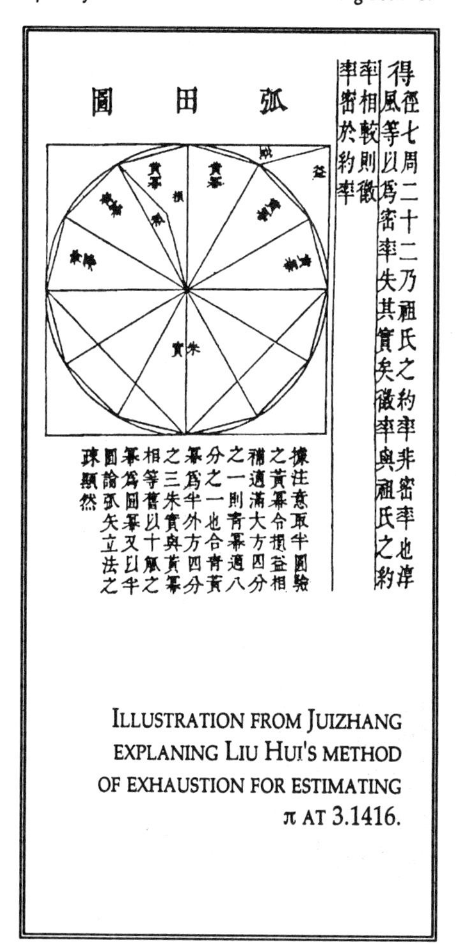

ILLUSTRATION FROM JUIZHANG EXPLANING LIU HUI'S METHOD OF EXHAUSTION FOR ESTIMATING π AT 3.1416.

Originally conceived as a reference work for bureaucratic officials, by the time of the Tang Dynasty (A.D. 618–907), the *Nine Chapters* had been incorporated into an official mathematics curriculum for the Chinese Empire. In turn, this curriculum was adopted by Korea and Japan; thus, officials throughout the East studied the methods and problem-solving techniques of the *Nine Chapters*.

The book is the oldest complete collection of applied problems in existence. As such, it provides insights as to how early people conceived of and used mathematics. However it is not solely the scope of the *Nine Chapters'* contents that makes it historically valuable but also the mathematics used. Let's briefly survey some of the mathematical highlights of this work.

Common fractions were employed extensively in computation. The Chinese represented these fractions as two vertically arranged numbers on the counting board; the upper number was the numerator termed "the son," the lower number, the denominator, "the mother." Popularly used fractions possessed special names: 1/2 "middle half," 1/3 "diminished half," 1/4 "weak half," etc. The *Nine Chapters* contains the first documented, systematic treatment of fractions in history. Fractions were reduced:

> *When capable to be halved, first halve them; if not capable, take the denominator and the numerator, and from the greater of the two subtract the other, and repeat the same operation until the greatest common factor is found, with which the fraction is to be reduced.*

Addition and Subtraction of Fractions

Addition and subtraction of fractions was accomplished with the use of a least common denominator. For example, problem 6, Chapter 1 requires the calculation:

$$1+\frac{1}{2}+\frac{1}{3}+\frac{1}{4}+\frac{1}{5}+\frac{1}{6}+\frac{1}{7}$$

which is undertaken:

$$\frac{420}{420}+\frac{210}{420}+\frac{140}{420}+\frac{105}{420}+\frac{84}{420}$$

$$+\frac{70}{420}+\frac{60}{420}=\frac{1089}{420}.$$

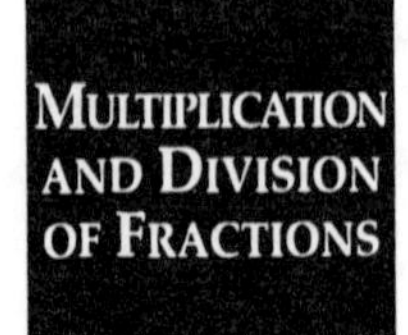

Multiplication and Division of Fractions

Multiplication and division of fractions conform to modern practices. For multiplication, the problem solver was instructed:

> *Take the product of the denominators as the divisor, that of the numerators as the dividend, and carry out the divisor.*

In division problems, a common denominator is found and the resulting algorithm as exemplified in modern notation is

$$\frac{a}{b}+\frac{c}{d}=\frac{ad}{bd}+\frac{bc}{bd}=\frac{ad}{bc}.$$

Previously, it was generally held that such a systematic discussion of fractional operations first occurred in India in the seventh century. Now it is known to have taken place in China much earlier.

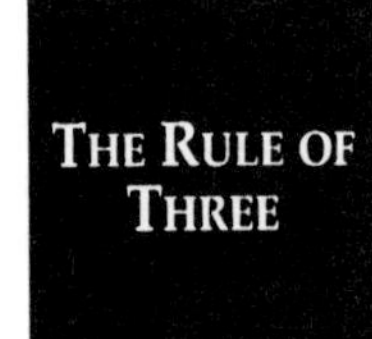

The Rule of Three

As demonstrated by the problems in Chapters 2 and 3, the ancient Chinese were proficient in using proportions and, in particular, in applying the "Rule of Three" in problem solving:

> *A man brings 12* jin *of gold through customs. The tax levied is one tenth of the value. If the customs officer takes as tax 2* jin *of gold and pays the man back 5000* qian *in cash, find the value of 1* jin *of gold. Ans: 6250* qian.

> *Now there are five persons of preceding degrees of rank. They have jointly hunted five deer. How many deer does each receive if all the deer are to be shared between them according to their precedence in rank? Ans: 1-2/3, 1-1/3, 1, 2/3, 1/3 (highest to lowest rank).*

"Construction Consultations," Chapter 5, reveals that a high level of spatial visualization existed in early China. Correct formulas for computing the volume for a variety of solids were given, for example:

Tetrahedron (Turtle's Shoulder Blade)

$$V=\frac{1}{6}a\,b\,h.$$

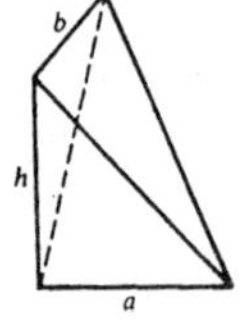

Wedge-Shaped Prism (Tomb Entrance Tunnel)

$$V=\frac{1}{6}(a+b+c)hl.$$

In Medieval Europe, the "Rule of False Position," whereby simple linear equations of the modern form $ax + b = 0$ could be solved, was considered a powerful, algebraic technique. Transmitted from the East through Arabic works, it was copied, renamed by European scholars such as Fibonacci, and promoted in the later works of Pacioli and Tartaglia. The Rule of False Position worked as follows:

> To solve $ax + b = 0$, make two guesses g_1, g_2 for the value of x then
>
> $ag_1 + b = f_1$ and $ag_2 + b = f_2$, where f_1 and f_2 are failures for the solution but using them we obtain
>
> $$x=\frac{f_1g_2-f_2g_1}{f_1-f_2}.$$

The Excess and Deficiency chapter of the *Nine Chapters* is completely devoted to applications of the Rule of False Position. Its problems truly challenge the ability of the reader, for example:

> *A man invests a certain amount of money in Sichuan, where the interest rate is 30%. He withdraws 14,000* qian *the first time, 13,000 the second time, 12,000 the third time, 11,000 the fourth time, and 10,000 the fifth time. On the last occasion, he has withdrawn all his money including interest. Find the amounts of his initial investment and his total interest. Ans: the initial investment is* $30468\frac{84876}{371293}$ qian *and the interest is* $29531\frac{286417}{371293}$ qian.

Now, it appears that the original source for the European version of the Rule of False Position was Cathay.

The Most Startling Chapter

The most startling of the *Nine Chapters'* mathematical revelations are found in Chapter 8. Here, by the use of rectangular arrays of counting rods on a counting board, techniques are described to solve systems of simultaneous equations as complex as five equations involving six unknowns, an indeterminate situation. A typical problem from this section with its solution scheme is as follows:

> *There are 3 bundles of top grade rice, 2 bundles of medium grade and 1 bundle of low grade yielding 39* dou*; 2 bundles of top grade, 3 bundles of medium grade, and 1 bundle of low grade yield 34* dou*; while 1 bundle of top grade, 2 bundles of medium grade and 3 bundles of low grade yield 26* dou*. Find the measure [of rice] contained in each bundle of the three grades of grain.*

In modern notation, the situation can be stated as:

$$3x + 2y + z = 39$$
$$2x + 3y + z = 34$$
$$x + 2y + 3z = 26.$$

The Chinese set up the following equivalent configuration on their counting board.

top grade	1	2	3
medium grade	2	3	2
low grade	3	1	1
yield	26	34	39

Designating the columns from right to left as C_1, C_2 and C_3, the following operations were then performed as in **Figure 1**.

Scholars in ancient China used matrix reduction techniques to solve systems of equations! Further, the general application of this method resulted in situations where a large column entry was subtracted from a smaller entry resulting in a negative number. The Chinese developed rules for such situations:

> *For subtracting: same sign take away, different signs add together, positive from nothing makes negative, negative from nothing makes positive;*

> *For addition: different signs take away, same signs add together, positive and nothing is positive, negative and nothing makes negative.*

In the computing processes, black rods represented negative numbers and red rods positive numbers. Problems were even initially stated with negative coefficients:

> *By selling 2 cows and 5 goats to buy 13 pigs, there is a surplus of 1000* qian*. The money obtained from selling 3 cows and 3 pigs is just enough to buy 0 goats. By selling 6 goats and 8 pigs to buy 5 cows, there is a deficit of 600* qian*. What is the price of a cow, a goat and a pig?*

These problems and their solutions prove that the Chinese were the first people to successfully use negative numbers in their mathematics!

The final chapter of the *Nine Chapters* presents a collection of twenty-four problems involving right triangle computations. It is the most complete collection of such problems from the ancient world and is available in English translation [Swetz & Kao, 1977]. The problems are ingeniously conceived, for example:

> *Two men starting from the same point begin walking in different directions. Their rates of travel are in the ratio 7:3. The slower man walks toward the east. His faster companion walks to the south 10* pu *and then turns toward the northeast and proceeds until both men meet. How many* pu *did each man walk? Ans: the faster 24 1/2* pu*; the slower 10 1/2* pu.

A complete English language translation of the *Nine Chapters of the Mathematical Art* should appear within the next five years. Then the full richness of this great mathematical work will be better enjoyed. ❑

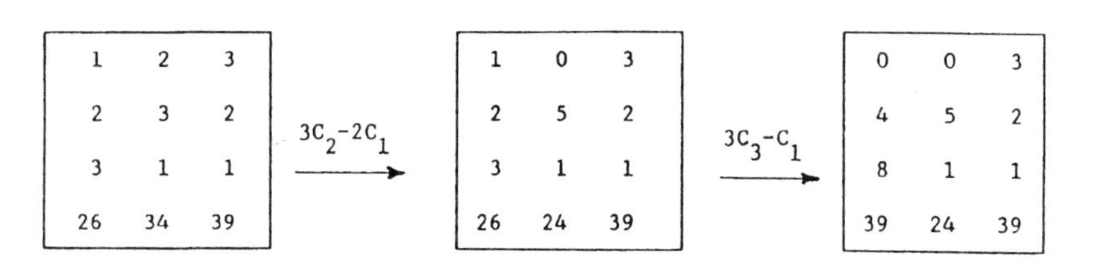

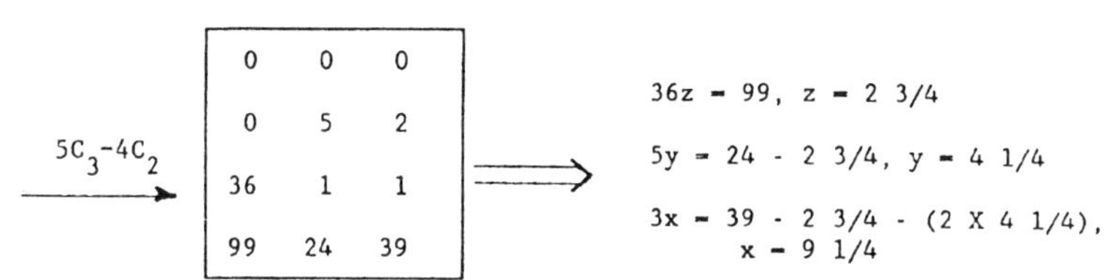

Figure 1.

Reference

Swetz, Frank and T. I. Kao. 1977. *Was Pythagoras Chinese? An Examination of Right Triangle Theory in Ancient China.* University Park, PA: The Pennsylvania State University Press.

Zero:
The Exceptional Number

Stephanos Gialamas
Miriam K. McCann

The history of zero has two basic branches: that meaning "empty space," and that which represents a number and is used for computation. Today, the same symbol and the same words are used for both situations (in the United States, at least). The use of one symbol for both is tied to the adoption of our modern number system.

In spoken and written language, the use of a zero is virtually unnecessary. For instance, the number 3052 is "three thousand fifty-two." The words indicate the value. It becomes necessary to use a zero when we wish to write symbols for the numbers and indicate value using a positional number system or when the number is zero. How did this come about?

The Babylonians provided us with some of the first recorded uses of symbols for numbers. Wedge-shaped symbols produced by a stylus pressed into clay tablets indicated a *sexagesimal* (base 60), positional system. That is, the position of the symbol within the number indicated its value. No symbol for indicating an empty position in a number was found on these tablets, which date back to over 4000 years ago (2300–1600 B.C.).

In the sexagesimal system, there are few occasions when the use of a symbol to indicate an empty position is needed, and the early Babylonians relied on the context to make clear the value of the number system as written.

About the sixth century B.C., a specific symbol that indicated the absence of a symbol began appearing within numerals. This new addition to the symbolic notation did not appear at the end of a numeral, so some confusion about the value could still occur. Astronomical tables in later years began to show the "missing" symbol at the end of numerals [Menninger 1969].

The Babylonians were not the only civilization to leave behind evidence of a positional system and some way to indicate a missing numeral. The Egyptians also used a symbolic system to indicate numerical values, but their system was not positional. It did not use or require the use of a zero to prevent misinterpreting the value. The Mayans of Central America and southern Mexico used a vigesimal (base 20) system, which was also positional. They employed a symbol for zero that looked roughly like a half-closed eye [Cajori, 1926]. This number system was being used during the first century A.D. and was found on early calendars [Kline 1953].

The Incas of Peru kept records of transactions by knots on cords, called *quipu*. This decimal (base 10) system used different types of knots as well as position of the knots to indicate value. A gap between sets of knots showed the absence of a value (position). The absence of a knot showed that the last (units) position was empty. Cords could be combined, and an additional cord was attached to indicate the sum. It was probably easy for the "reader" to tell when a missing number was indicated. Cords with no knots could also indicate zero [Ascher & Ascher 1981].

Native North American people used numeration systems of various bases (3, 4, 5, 8, 10, and 20), but there is no indication of place values or the use of a zero [Boyer 1944].

Ptolemy (A.D. 130) was using a missing numeral symbol in tables of cords [Boyer 1968] which resembled the Greek letter *omicron*. Ptolemy apparently used the $\bar{o}$ symbol only with the sexagesimal fraction notation and not as part of the alphabetic numeral system of the Greeks. [Heath 1981] $\bar{o}$ was an abbreviation of the Greek word *ouden* or *outhen* meaning "nothing" [Menninger 1969].

The zero that represents a number and is used as such first appeared with the Hindu civilization in India. The earliest recorded evidence found so far is on the Gvalior inscription, found on the wall of a temple near Lashkar in Central India. The inscription gives the date 933 (our A.D. 870) and lists gifts to a temple. The numbers "270" and "50" appear, using a small circle for zero [Menninger 1969]. This zero was probably in use long before recorded evidence shows.

The mathematicians of the civilizations mentioned so far were aware of the concept of zero and references were made in their writings, but they did not have an established symbol representing the concept. Since today in mathematics we use a large number of symbols, it is hard to comprehend that the early mathematicians (actually until the Dark Ages) were limited in expressing their concepts in literature.

Sunya, in Sanscrit (sixth to eighth centuries A.D.) meant "empty" and became the Arabic *as-sifr* ("the empty") as the Hindu digits began spreading westward in the ninth century. During the thirteenth century A.D., two Latin words, *cifra* and *zefirum* (or *cephirum*) were used to mean zero. They were adapted into Latin from Arabic as the numbers became known in Europe [Menninger 1969]. These Latin words later became *cipher* and *zero*. The confusion created by a single word (cipher) meaning a number and also a digit which had numerical value, or of two words (cipher and zero), which each meant the same thing, helps to explain why it seemed to take so long for the new number system to become adopted.

Zero was also still used as a placeholder (the ancient notion of "empty" position) as well as a number in calculations. The dual role of zero, however, was not an easy concept to accept.

Early computations were done on various forms of counting boards, which existed in many civilizations. Some counting devices, such as the various forms of the abacus, are still being used today, while others are only known to us because they were illustrated on such items as vases, tombstones, or other artifacts [Menninger 1969].

Sand tables were used prior to the ninth century for computations on a flat surface covered with a light coating of sand. Marks in the sand indicated number values; computations were completed and the sand smoothed to begin again. Other computational devices included counting boards divided into columns or rows for place value, with various types of markers used to indicate value. The markers were moved on the board as the computations were made [Dantzig 1959] and the results of the computations were recorded on something more permanent. A form of zero was used on these devices to indicate an empty cell, column, or row.

PTOLEMY USED THIS SYMBOL ONLY WITH SEXAGESIMAL FRACTION NOTATION.

From India, the concept of zero spread to China and was incorporated into an existing rod system of numbers [Boyer 1968]. The rod numbers were positional and decimal, but there is no indication that they were used in computation. The Chinese also used a named place-value system using Chinese characters for written numbers, which did not require the use of a zero.

Hindu numerals spread to Europe as early as A.D. 1000 when Gerbert (who became Pope Sylvester II) used the digits learned from the Arabs on counters of a counting board. The digits were not accepted then because it was not understood that they could be used for computations without a counting board. Hence, their usefulness was not appreciated. The Hindu numbers, including zero, gradually gained acceptance in Europe through their use by merchants and tradesmen. Texts written by arithmeticians and calculators also contributed to the spread of the new system [Menninger 1969].

The invention of the printing press in the fifteenth century expanded Hindu numerals to more of the population as printed matter became more accessible and available. Some forces, however, helped to slow the spread, including the social structure, which kept written material from the lower classes. Counting devices, when used by those who were adept, were speedy and accurate, and therefore, made a written numeral system somewhat obsolete.

What one can recognize as a zero today did not come into common use until much later in history. Besides the Babylonian, Mayan, and Greek symbols, a few other symbols were used prior to the "0" of modern times.

The Hindu symbol was first a dot and later a "0" crossed by a horizontal or slanting line. In the translations of Al-Khowarizmi's astronomical tables, three signs for zero appear (*j*, *O*, and *t*—an abbreviation of *teca* or *theca*). These symbols were used periodically by various writers over the years [Cajori 1928]. Even today, zero is

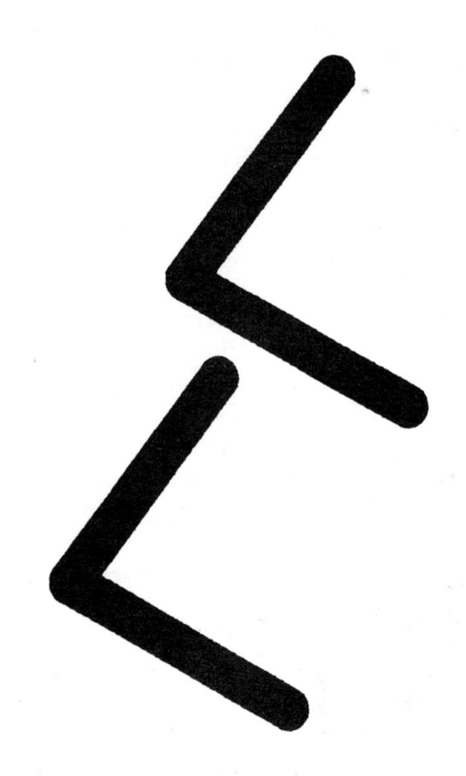

THE BABYLONIANS WERE PROBABLY THE FIRST TO USE A SYMBOL REPRESENTING AN EMPTY SPACE.

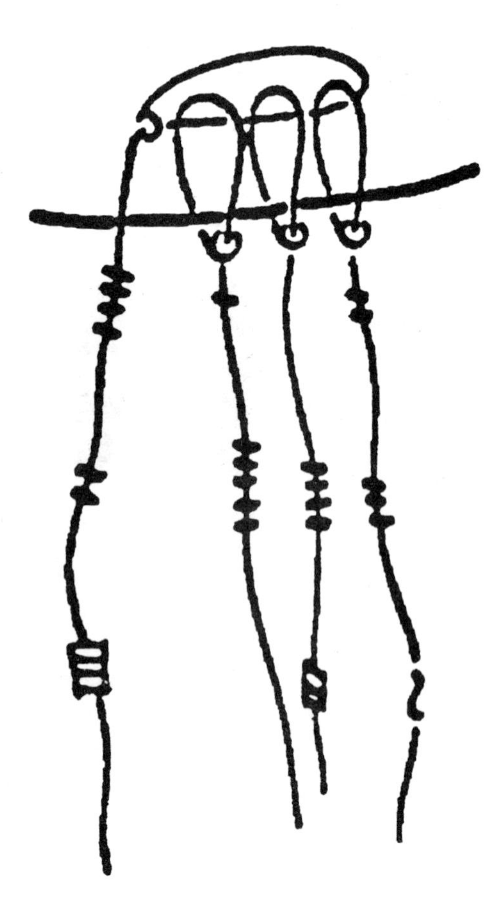

THE INCAS OF PERU USED KNOTS, CALLED *QUIPU* TO PORTRAY A BASE 10 SYSTEM.

occasionally written with the slash through it to distinguish it from the letter O (particularly in computer applications).

The fact that computations with zero seemed to have some strange results did not help people to understand the number. The difficulties with understanding the results of some calculations with zero were more evident as the new numbers were being assimilated into Western Europe.

In the ninth century, Mahavira wrote that a number divided by zero remains unchanged [Kline 1972]. Ehaskara (A.D. 1114) felt that a fraction with a denominator of zero remained the same even when anything was added to or subtracted from it. He also stated that a number divided by zero is an infinite quantity [Kline 1972]. These ideas were undoubtedly difficult to comprehend for those who were unfamiliar with computations using the 10 digits. If computational results could not be explained in physical terms, the results did not have meaning.

Subtraction of a number from zero was a particularly difficult concept to grasp. The resulting negative number was incomprehensible to some. Pascal (sixteenth century A.D.) regarded the subtraction of four from zero as "utter nonsense" [Kline 1972]. As late as the nineteenth century, Augustus DeMorgan felt that a negative number as a solution to an equation was "inconsistent, absurd" and had no "real meaning" [Kline 1972].

One fifteenth century French writer felt that zero was the creator of confusion and difficulties [Menninger 1969]. He noted that zero in front of a number (e.g., 03) did not affect the value of the number but zero behind the number multiplies it by 10 (e.g., 30).

THE CHINESE REPRESENTATION OF THE NUMBER 1,405,526.

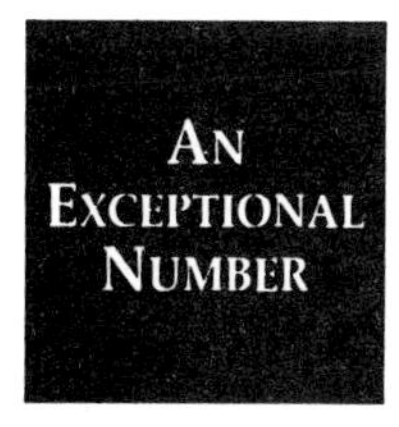

It is the number zero and its unique computational "rules" that seem to cause the most problems. In order for students to use the number properly, they must have some understanding of more advanced mathematical concepts. The first exposure is at the elementary level with zero as "nothing," which seems adequate for the young child, but which does not aid understanding when division using zero is encountered. In this case, it is important to understand the idea of inverse relationships in mathematics (specifically, that multiplication is the inverse operation for division). Without this understanding of mathematical structure, students will continue to have difficulty understanding division by zero.

Is the learning of both numbers and the concept of zero by children parallel to the development of the number system? Some aspects of the process seem so. Young children learn to count verbally starting with the number "one" just as our ancient ancestors did. Their early experiences with numbers are associated with the number of items in a set of objects, called *numerosity*. This first exposure does not include zero; however, the concept of an empty set is not a difficult idea and young children are quite aware when all of the items in question are gone. Pre-schoolers are also able to add and subtract, as long as the operations are performed on sets of objects. They are even able to do this in a verbally presented hypothetical situation, as long as the numbers are small (1, 2, 3) and the numbers being used are associated with items [Hughes 1986].

> "The confusion created by a single word ... meaning a number and also a digit which had numerical value, or of two words ... which each meant the same thing, helps to explain why it seemed to take so long for the new number system to become adopted."

For instance, if a child is asked, "What does 1 and 2 make?" he does not usually know. However, "What does 1 block and 2 blocks make?" elicits the correct answer. The pre-school child even accepts the removal (subtraction) of all of the items in the set resulting in a cardinality of zero.

These early childhood experiences remind us of the early civilizations who were counting and calculating using objects. What is surprising to many is that it took many centuries for people to adopt the ten-digit numerical system for computational purposes. A series of studies reported by Martin Hughes [1986] with pre-school- and early-primary-aged children in England sheds some light on this difficulty.

The children in Hughes' study were presented with 4 "tins" containing "bricks." Each tin contained 0, 1, 2, or 3 bricks. After the child saw and counted the number in each, the tins were covered and moved, and the child was asked to tell how many were in each tin. Since the tins were identical, correct answers were purely chance. Then, the children were told they could give themselves hints about the numerosity of the contents by marking the paper attached to the top of each tin. The type of mark was entirely up to the child [Hughes 1986].

Successful students, those whose marks allowed them to identify the contents of the tin immediately after making the

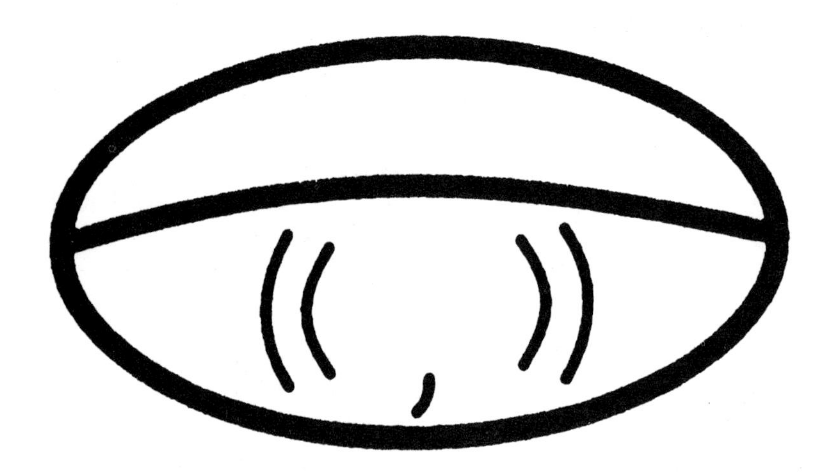

THE MAYAN "EMPTY POSITION."

THE MEDIEVAL NAME FOR ZERO APPEARS AS *THECA* IN THIS 15TH CENTURY MANUSCRIPT.

marks as well as at a later time, used either the symbols 0, 1, 2, 3, or tally marks, or the same number of some created symbol or picture. In a similar study, the students were given magnetic numbers to use on the tins. Some students responded by placing the appropriate number of pieces on the tin rather than the appropriate number symbol [Hughes 1986]. When the tester showed the students how they could indicate the number in each tin, they caught on quickly and were successful in marking the tins and identifying the contents from their indications.

The same type of game was played where the number of bricks was increased or decreased, and the children were to indicate what had happened. Only the students with the greater mathematical ability used formal symbols to indicate the operation when initially introduced to the game [Hughes 1986].

Children in both studies indicated that the empty tin had nothing in it either by placing nothing on the tin or using the symbol 0. The students who used the number of pieces or tally to indicate the numerosity left the top of the tin blank, while those who used the number symbols to indicate the count used the zero to indicate the empty tin [Hughes 1986].

These studies with children indicated that they did not pick up the ideas that the counting, adding, and subtracting they were doing on paper in their math classes had any connection to the counting, adding, and subtracting that they could do with objects. With instruction, however, the students did see the connection and were successful in future games.

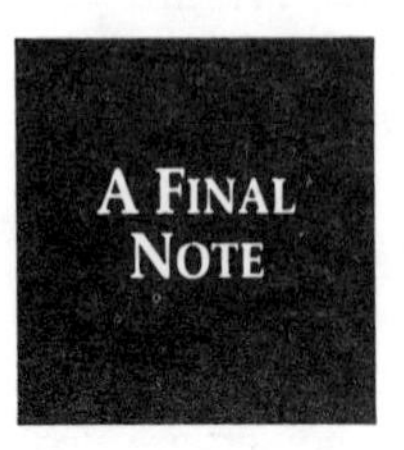

The same lack of understanding seemed to be part of the reason the Hindu system of numbers and calculations was so long in being accepted. The use of calculating devices was widespread, and the objects on the boards represented sets of things; the abstract computations required a whole new way of dealing with numbers.

Just as children need to be taught how to relate counting and the concept of zero as a number, so did the people in our history. Until printing enabled the skill of computation to spread quickly and people opened their minds to something new, the spread of the Hindu numbers and zero was bound to be slow. ❑

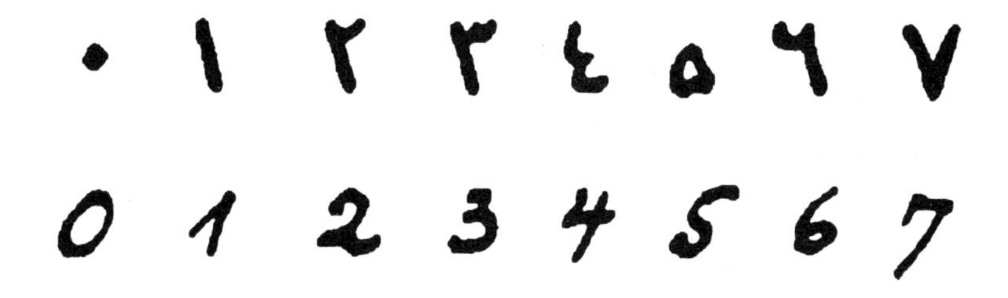

THE EAST ARABIC NUMERALS (TOP) EVOLVED INTO OUR MODERN NUMERALS (BOTTOM).

References

Ascher, Marcia and Ascher, Robert. 1981. *Code of the Quipu*. Ann Arbor, MI: University of Michigan Press.

Boyer, Carl B. 1944. Zero: the Symbol, the Concept, the Number. *National Mathematics Magazine*. 18:323–330.

Boyer, Carl B. 1968. *A History of Mathematics*. New York: John Wiley & Sons, Inc.

Cajori, Florian. 1926. *A History of Mathematics*. New York: The Macmillan Co.

Cajori, Florian. 1928. *A History of Mathematical Notations, Vol. I*. LaSalle, IL: The Open Court Publishing Company.

Dantzig, Tobias. 1959. *NUMBER: The Language of Science*. 4th ed. New York: The Macmillan Co.

Hughes, Martin. 1986. *Children and Numbers*. New York: Basil Blackwell, Inc.

Kline, Morris. 1953. *Mathematics in Western Culture*. New York: Oxford University Press.

Kline, Morris. 1972. *Mathematical Thought from Ancient to Modern Times*. New York: Oxford University Press.

Menninger, Karl. 1969. *Number Words and Number Symbols: A Cultural History of Numbers*. Cambridge, MA: M.I.T. Press.

THE FIGURES IN THIS ARTICLE ARE FROM KARL MENNINGER'S *NUMBER WORDS AND NUMBER SYMBOLS: A CULTURAL HISTORY OF NUMBERS*. CAMBRIDGE, MA: THE M.I.T. PRESS, 1969.

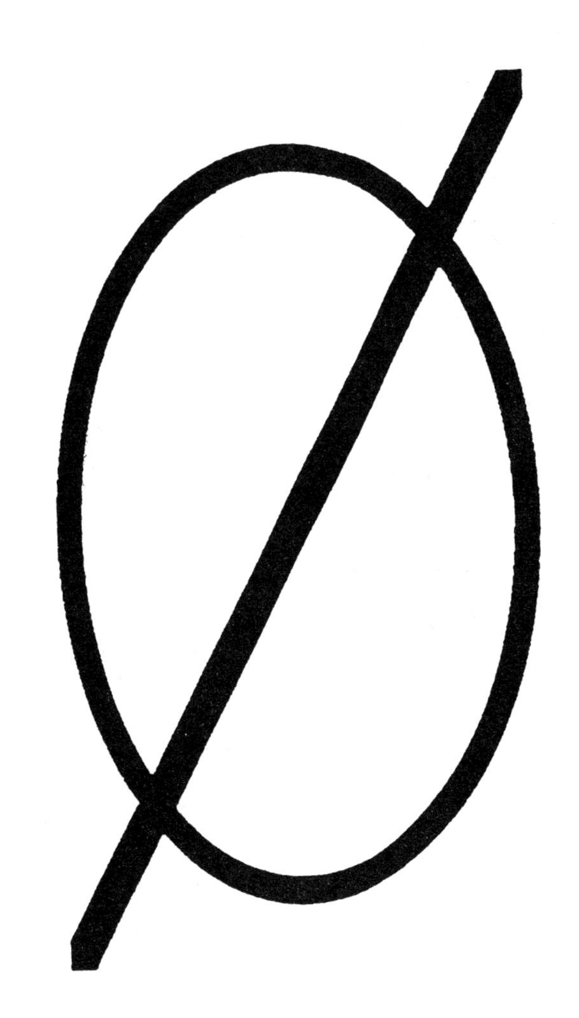

A COMPUTER USES THE "NULL" SIGN TO DISTINGUISH ZERO FROM THE LETTER "O."

A Madness in the Methods

Cubic and Quartic Equations: Are the General Solving Techniques Still Important?

RICHARD L. FRANCIS

An exciting part of the history of mathematics is the sixteenth century discovery of general methods for solving both cubic and quartic equations. Some traces of limited success with such higher degree equations can be identified in more ancient times. Yet, it remained for Italian mathematicians of the late pre-modern era to resolve solving of these equations completely.

What, however, were these methods stemming from a certain mathematical madness? With whose names in history are the methods today associated? And how do they relate to what is called a fundamental theorem of algebra? Student encounters with things mathematical parallel closely mankind's historical contact; consider thus a challenging classroom encounter with a remarkable part of the history of mathematics, namely, the general equations of the third and of the fourth degree.

An essential part of algebra is the solving of the general quadratic equation. Although certain cubic equations, such as $x^3 - 3x^2 + 2x = 0$, or quartic equations, such as $x^4 - 10x^2 + 9 = 0$, are discussed, more general cubic and quartic types are omitted. Not all students will express curiosity as to whether solving techniques exist for these higher degree equations. Yet, occasionally, the perceptive student will pursue this line of questioning.

The methods used to solve such equations are indeed within the capabilities of some students in an early phase of mathematical training. Moreover, such methods offer a promising, fruitful field of endeavor, resulting in both a deeper understanding and enhanced applications.

Solving of cubic and quartic equations seems to have its greatest appeal at the time of the student's encounter with restricted higher degree equations. Prior to explaining these methods, lessons should include a detailed study of quadratic equations and complex numbers. These should be supplemented by such skills as synthetic division and rational root analysis. An important, specific item concerns the student's awareness that the three cube roots of 1 are given by the numbers 1, $(-1 - i\sqrt{3})/2$, and $(-1 + i\sqrt{3})/2$. Respective representation of the imaginary roots by w and w^2 is an easily verified relationship; it also follows that $w^3 = 1$. (*Note:* solving the special cubic equations $x^3 - 1 = 0$ by factoring yields the listed cube roots of 1.)

How might such a study begin? To gain perspective, the general equation type and any desired restrictions should be stressed. Using a rather clever subscript–exponent pattern, this equation may be written as,

$$a_0x^n + a_1x^{n-1} + a_2x^{n-2} + \dots + a_{n-1}x^1 + a_n = 0.$$

Classification of the above is needed: if $a_0 \neq 0$, the equation is of degree n, and by the Fundamental Theorem of Algebra, has at least one solution in the field of complex numbers [C.F. Gauss (1777–1855)]. It can further be established that the equation has n solutions, though these solutions need not be distinct. The statement regarding the number of solutions is true even if the coefficients are restricted to the set of real numbers. Such a restriction will be applied throughout this article. As a consequence of the coefficient restriction, imaginary roots will occur in conjugate pairs.

If n is set equal to 3 in the preceding equation, a general cubic results. Such an equation is often written in the form,

$$ax^3 + bx^2 + cx + d = 0.$$

As imaginary roots occur in conjugate pairs, the equation will have exactly one real solution, or all of its solutions will be real numbers. This latter case gives rise to a trigonometric approach in solving (often called the *irreducible case*), and will not be pursued here because of our essential restriction to algebra.

It is easy to verify that if $x = y - (b/3a)$ in the above equation, the second-degree term will disappear. In other words, a more easily manipulated cubic results.

It reads as follows:

$$Ay^3 + Cy + D = 0.$$

The method of solving a general cubic equation is at best explained by an illustration. One can quickly verify that the equation below has no rational roots; consider,

$$x^3 + 6x^2 + 6x + 2 = 0.$$

If $x = y - (b/3a)$, $x = y - 2$, then $y^3 - 6y + 6 = 0$. At this point, let $y = r + s$, so as to obtain,

$$r^3 + s^3 + (r + s)(3rs - 6) + 6 = 0.$$

The middle portion of the equation can always be factored so as to have $r + s$ as a factor. Now impose a condition on the remaining factor by setting $3rs - 6 = 0$; this in no way affects the solution set. Then $rs = 2$ or $r = 2/s$. The equation above becomes by this substitution,

$$s^6 + 6s^3 + 8 = 0,$$

and here, as in the general case, is a quadratic in s^3. Solving (by the quadratic formula or by factoring),

$$(s^3 + 2)(s^3 + 4) = 0.$$

Using either real value of s, say $s = -\sqrt[3]{2}$ r becomes $-\sqrt[3]{4}$. Now $rs = 2$ is also satisfied by $(rw)(sw^2) = 2$ and by $(rw^2)(sw) = 2$; recall here that $w^3 = 1$. So the y solutions are given by,

$$y = r + s = -\sqrt[3]{2} - \sqrt[3]{4}$$
$$\text{or } y = rw + sw^2 = -w\sqrt[3]{2} - w^2\sqrt[3]{4}$$
$$\text{or } y = rw^2 + sw = -w^2\sqrt[3]{2} - w\sqrt[3]{4}.$$

As $x = y - 2$, the three x-values constituting the solution set are quickly written.

The method used above for solving cubic equations is called Cardano's Method (Geronimo Cardano, 1501–1576; also called G. Cardan in some sources); by some, it is called the *Cardano–Tartaglia Method.* Such a technique dates from the sixteenth century.

Rather frequently, solutions to the general cubic equation are given according to a formula, where one simply substitutes to obtain results. Obviously, this hardly answers the question as to how the formula was obtained in the first place. The ideal is to be able to solve the various cubics without having to resort to the formula in each and every case; mastery of the detailed process gives evidence of a deeper understanding.

If Cardano's Method is to be used to advantage in teaching, the coefficients ought to be carefully selected. This circumvents the students' being sidetracked on unduly lengthy computations.

A disposition of the cubic problem makes one consider the next equation type, namely the quartic or bi–quadratic. The general quartic equation is often written as,

$$ax^4 + bx^3 + cx^2 + dx + e = 0,$$

where $a \neq 0$. Basically speaking, its solution rests on solving a cubic equation, and that in turn rests on solving a quadratic equation. The method fundamentally calls for setting one square expression equal to another such expression.

Consider the following:

$$x^4 - 8x^3 + 20x^2 - 16x - 21 = 0.$$

First write the third- and fourth-degree terms on the left, and then complete the square of the left-hand side. The given equation then becomes,

$$x^4 - 8x^3 = -20x^2 + 16x + 21$$
$$\text{or } x^4 - 8x^3 + 16x^2 = -4x^2 + 16x + 21$$
$$\text{or } (x^2 - 4x)^2 = -4x^2 + 16x + 21.$$

As the left-hand side is a square, it is fairly simple to introduce another quantity, y, so as to equate one square expression to another. Here, add $(x^2 - 4x)y + y^2/4$ to each side of the last equation. The following results:

$$(x^2 - 4x)^2 + (x^2 - 4x)y + y^2/4$$
$$= (-4x^2 + 16x + 21) + (x^2 - 4x)y + y^2/4.$$

The left-hand side in this equation is also a square (an intended construction); the right-hand side will become a square provided its discriminant is zero. Rearranging, the right-hand side becomes,

$$x^2(-4 + y) + x(16 - 4y) + (y^2/4 + 21).$$

Setting its discriminant equal to zero gives the equation,

$$(16 - 4y)^2 - 4(-4 + y)(y^2/4 + 21) = 0,$$
$$\text{or } y^3 - 20y^2 + 212y - 592 = 0.$$

Fortunately, this cubic (called the *resolvent cubic*) is solvable by the rational root approach; otherwise Cardano's Method would be required. A rational solution is given by $y = 4$; actually any one of the three solutions will suffice in completing the problem. The original equation in both x and y becomes,

$$[(x^2 - 4x) + y/2]^2 = (-4x^2 + 16x + 21)$$
$$+(x^2 - 4x)y + y^2/4$$
$$\text{or, } (x^2 - 4x + 2)^2 = 25.$$

Solving the equations $x^2 - 4x + 2 = 5$ and $x^2 - 4x + 2 = -5$, the x-values of $2 \pm \sqrt{7}$ and $2 \pm i\sqrt{3}$ result.

In applying this general method for solving quartics, the right-hand side may well turn out to be a square expression in x, as opposed to a numerical quantity, as happened here. Whichever occurs creates no critical problem; the quadratic formula suffices in either case. Again, a careful selection of coefficients is suggested if the method is to be used effectively in the classroom.

This technique of solving a quartic equation, too, stems from the sixteenth century, a time of considerable mathematical activity in the development of a theory of equations. It is often called *Ferrari's Method* (L. Ferrari, 1522–1565), after a pupil and collaborator of Cardano's. Cardano later published the formula in his *Ars Magna* [Gellert, et al. 1975].

The remaining higher degree equations (quintic, sextic, etc.) pose a more difficult problem than those just considered. They cannot be solved radically in terms of their coefficients. Such a discovery is a notable part in the unfolding story of mathematical achievements, and involves an impressive list of mathematicians (P. Ruffini, N.H. Abel, E. Galois, and others).

As mentioned, not all students can handle these complex problems. But, those who are somewhat more advanced and curious will benefit from these lessons, and several outcomes can be expected, including:

Piracy?

"...Niccolo Tartaglia (1500–1557), mathematician and ballistic engineer, had found the formula [to the cubic equation], which today is named after Cardano, and had achieved considerable fame by applying it brilliantly in public problem-solving contests which were customary at the time."

[Gellert, et al., 1975]

or Privilege?

"The ambitious Professor Geronimo Cardano (1501–1576) of Venice, who did not succeed in finding the formula to the solution, obtained it from Tartaglia in 1539, after years of intense pressure, but he had to swear a solemn oath to Tartaglia to treat it as a kind of professional secret. However, Cardano broke his promise and included the result in his *Artis magnae sive de regulis algebraicis* (that is, of the Great Art of the Rules of Algebra) of 1545. . . .since the formula appeared in print . . . under Cardano's name, it became known as Cardano's formula. Even Tartaglia's protest, which led to a violent quarrel, was of no avail."

[Gellert, et al., 1975]

1. The writing of exact solutions, as opposed to interpolative approximations of real roots.

2. The obtaining of all imaginary solutions, which is generally impossible in a restricted discussion of real root approximations.

3. The broadening of the field of applications, wherein cubic and quartic models or set-ups are now within solving reach of the student.

4. The forming of a firmer basis for advanced pursuits, thus giving the student an enhanced insight and perspective.

5. The acquiring of a feeling of independence, a desirable outcome in that the student need not be overly reliant on an assortment of formulas.

6. The appreciation that stems from an awareness of important landmarks in the history of mathematics.

Admittedly, the inclusion of such material from the theory of equations is a challenging one. Yet to the perceptive student who does the usual work of the class with minimal effort, it may well prove a promising field of study. ❑

References

Francis, Richard L. April 1986. Did Gauss Discover That Too? *Mathematics Teacher* 79: 226–230.

Gellert, W., H. Küstner, M. Hellwich, and H. Kästner. 1975. *The VNR Concise Encyclopedia of Mathematics*. New York: Van Nostrand Reinhold Company.

The Lost Chord?

RICHARD L. FRANCIS

Formalized trigonometry likely originated with the works of Hipparchus around 140 B.C. Its primary emphasis in the long-ago setting was that of select arcs and associated chords which enabled the prober to solve astronomical triangles. Such a literal meaning of trigonometry (that of "triangle measure") thoroughly typified the discipline until its modern-era development. Because of the non-analytical features of early trigonometry, functional values of extremely small angles were beyond the bounds of evaluation.

CARL FRIEDRICH GAUSS (1777–1855)

It was long believed that the ancient Greeks had closed the subject of regular polygon construction. Yet, their contributions were restricted to regular polygons having three, four, five, and fifteen sides, along with certain multiples of these. Minimal efforts were made over the centuries to find construction techniques, using only the Euclidean instruments of the unmarked straightedge and the compass, for regular polygons of seven, nine, eleven, and thirteen sides. None proved successful. However, in 1796, a breakthrough occurred in the characterization of regular polygon constructibility.

At barely 19 years of age, the German mathematician Carl Friedrich Gauss (1777–1855) discovered and proved that a regular polygon of seventeen sides could be inscribed in a circle by means of a compass and straightedge alone. Gauss was so excited about his discovery that he requested that the figure be inscribed on his tombstone!

The extended proof of this discovery consisted of two parts. The first of these established that a prime-sided polygon is constructible if and only if the number of sides is of the form $2^n + 1$, where $2^n + 1$ is prime. Rather simply, such a prime can be written in the form $2^{2^r} + 1$, where r is, at this time, equal to 0, 1, 2, 3, or 4. Any prime of this form is called a *Fermat prime* (Pierre de Fermat, 1601–1665).

The five Fermat primes shown in **Table 1** are the only ones presently known. However, on the basis of this listing alone, three additional constructible regular polygons have been added to those known by the Greeks. Although Gauss's accomplishment was to say which of the prime-sided regular polygons are constructible, only in the case for the the regular *heptadecagon* (seventeen-sided figure) did he give an actual detailing of the construction steps.

Fermat Number Table

r	$2^{2^r} + 1$	Primality of $2^{2^r} + 1$
0	3	yes
1	5	yes
2	17	yes
3	257	yes
4	65537	yes

TABLE 1.

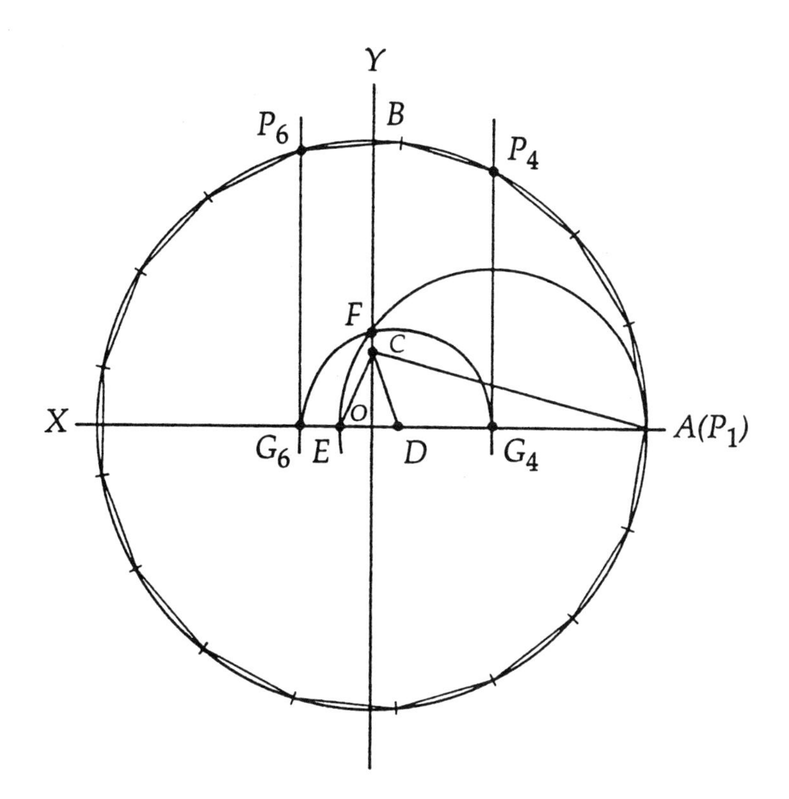

Figure 1. A regular heptadecagon inscribed in a circle.

To construct a polygon of seventeen sides, let OA and OB be two perpendicular radii of a given circle with center O. Find C on OB such that $OC = OB/4$. Now find D on OA such that $\angle OCD = (\angle OCA)/4$. Next find E on AO produced such that $\angle DCE = 45°$. Draw the circle on AE as diameter, cutting OB in F, and then draw the circle $D(F)$, cutting OA and AO produced in G_4 and G_6. Erect perpendiculars to OA at G_4 and G_6, cutting the given circle in P_4 and P_6. These last points are the fourth and sixth vertices of the regular heptadecagon whose first vertex is A (**Figure 1**) [Eves 1963].

By Gauss's prime standard, many of the long-sought Greek constructions became classifiable. In particular, the regular heptagon (seven-sided polygon) was given the distinction of being the first of the non-constructible regular polygons. This is clear since 7 is a prime not of the form $2^n + 1$. Similar arguments apply to regular polygons of sides numbering 11, 13, 19, and other primes which deviate from the form $2^n + 1$.

Primes generated by the expression $2^n + 1$ were of interest to Fermat and others well before their application in geometry. Fermat believed the set of such primes was infinite according to the conjecture that $2^n + 1$ is prime whenever n is a power of 2. Although the primality of $2^n + 1$ indeed implies that n is a power of 2, the converse was shown false by Leonhard Euler (1707–1783) in 1732, nearly a century after Fermat. Euler's counterexample was the number $2^{2^5} + 1$. In showing that $2^{2^5} + 1$ had (641)(6700417) for its factorization, Fermat's guess was thus proved incorrect.

Such a breakthrough highlighted the number theory problem in the years prior to Gauss by focusing attention on the inquiry, "What then is the cardinality of the set of Fermat primes?" Not even Gauss supplied an answer to this question; it is still unanswered today.

The Gaussian Criterion

Not only did Gauss and his contemporaries answer the construction question for regular polygons of a prime number of sides, but also for all regular polygons, whatever the number of sides. The key form in disposing of the famous constructibility problem is actually the product of a power of 2 and a collection of distinct Fermat primes. Such a form, called a *Fermat product*, can be written concisely as

$$(2^k)(p_1)(p_2)(p_3) \ldots (p_j),$$

where k is a non-negative integer and the p's denote distinct Fermat primes. Very simply then, a regular polygon of x sides is constructible if and only if x is a Fermat product or a power of 2 greater than or equal to 4.

Gauss's wish that his tombstone show a regular heptadecagon inscribed in a circle proved to be quite difficult. The monument, found today in Braunschweig (Brunswick), Germany, the place of Gauss's birth, shows a seventeen-pointed star—not the figure he requested. Oddly enough, the stonemason refused to chisel out the heptadecagon because he believed it would look like a perfect circle. Was the stonemason correct?

If the monument, and hence the circle, had been larger, the task might have been looked upon more favorably. But either way, the circular arc would scarcely, if at all, prove visibly distinguishable from the polygon that it circumscribes.

Again, we know that a regular polygon of an odd number of sides x is constructible if and only if x is a prime of the form $2^n + 1$ or the product of distinct primes all of this form.

The Gaussian epitaph depended on visibly distinguishing an arc from the side of a polygon. To quantify the subjective term "distinguishable," let's first theorize and ignore all thickness in the practical situations of geometric drawings and inscriptions. Let's also agree that a chord of a regular polygon is quasi-distinguishable from the associated arc of the circle if and only if the circular segment has an area of at least one square millimeter.

All of this raises the question, how might the epitaph problem unfold if other side numbers are considered? Such a question presents an assortment of complexities in the setting of high precision. It is also an interesting twist on a bit of mathematical history.

Pursuit of "practically negligible" values often rules out conventional approaches. The high degree of precision implied necessitates more than the use of readily available tables and quick computations with pencil and paper. Reinforcement of this point of precision is nicely provided by a variation on the Gaussian epitaph problem. The chord it suggests is all but lost in a maze of scarcely distinguishable areas.

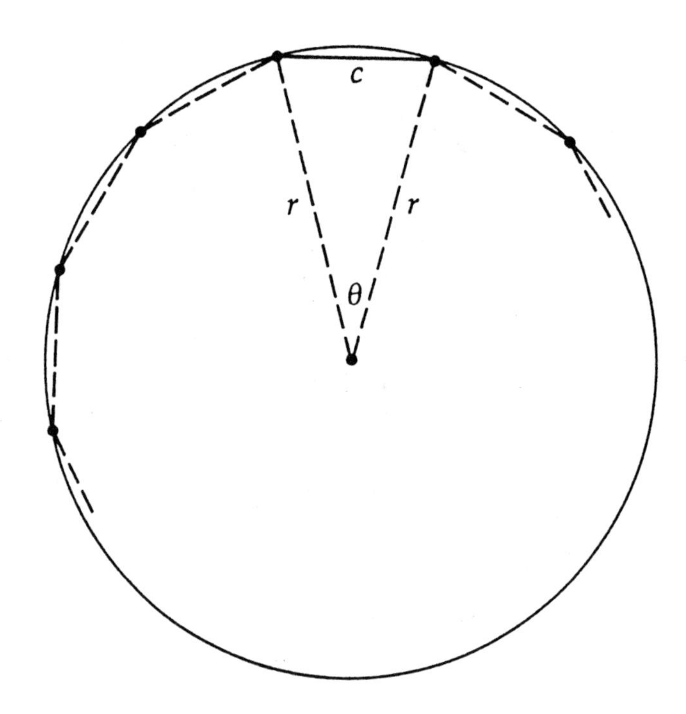

FIGURE 2. THE SIDE OF A REGULAR INSCRIBED POLYGON.

The maximum odd number of sides of a regular polygon that is known to be constructible is given by

$$(3)(5)(17)(257)(65{,}537).$$

These numbers include all known primes of the form $2^n + 1$. This large number is actually 4,294,967,295.

Suppose the problem of Gauss's epitaph is changed from seventeen sides to that of the gargantuan number above. What must the radius of a circle be so that a chord of this circle is quasi-distinguishable from the associated arc if indeed a regular polygon of 4,294,967,295 sides is inscribed? Before elaborating on the computational refinements involved, we will first note the various numerical outcomes.

The central angle of a regular polygon of 4,294,967,295 sides is

$$\theta = \frac{2\pi}{4{,}294{,}967{,}295} \text{ radians.}$$

Hence, θ is approximately $1.462918079607772(10)^{-9}$ radians. The area of the circular segment associated with this follows from the formula $A = 1/2r^2(\theta - \sin\theta)$. As the value of $(\theta - \sin\theta)$ is approximately $5.218056428083056(10)^{-28}$, the formula above reduces to

$$1 = {}^1/_2\, r^2[5.218056428083056(10)^{-28}].$$

Solving for r, an approximate value of $6.190997265735683(10)^{13}$ millimeters is obtained. The millimeter has universal overtones; it stems from the definition of the meter, which is traditionally one ten-millionth of the length of the earth's quadrant (e.g., one ten-millionth of the distance of the North Pole from the Equator).

An arc of the circle above would appear deceptively straight and truly non-distinguishable from the associated chord. By our definition however, and microscopically so, the label "quasi-distinguishable" will apply. It is quickly noted that r is roughly $6.2(10)^7$ kilometers. This circle, having a radius of about 39,000,000 miles, is of course mind-boggling. A great circle of the sun would be small in comparison to it.

The chord which is essentially lost in this discussion is the side of the inscribed polygon (see **Figure 2**). Such an elusive, hard-to-envision chord c can be "found" by use of the Law of Sines or the Law of Cosines. Using the latter, note that

$$c^2 = 2r^2 - 2r^2\cos\theta \text{ or}$$
$$c^2 = 2r^2(1 - \cos\theta).$$

That is, $c = r\sqrt{2(1 - \cos\theta)}$. In this case, $\cos\theta$ is very close to 1 in value as the positive θ-value is nearly zero. The radius, though enormous, is multiplied by an extremely small number. What then is the chord's length? Roughly 0.0905692183084701 kilometers. This translates into 0.05627708574955 miles or roughly 297.14 feet. Note that $\sin\theta$ and $\cos\theta$ are related concisely by the identity $\sin^2\theta + \cos^2\theta = 1$.

Although the formula $A = {}^1/_2\, r^2(\theta - \sin\theta)$ seems simple enough, various challenges are attached to finding θ as well as $\sin\theta$ in order to make a substitution possible. Here pin-pointed is the inadequacy of the conventional or standard trigonometric approach.

Gauss's wish that his tombstone show a regular heptadecagon inscribed in a circle proved to be quite difficult. The monument, found today in Braunschweig (Brunswick), Germany, the place of Gauss's birth, shows a seventeen-pointed star—not the figure he requested. Oddly enough, the stonemason refused to chisel out the heptadecagon because he believed it would look like a perfect circle. Was the stonemason correct?

Beyond Conventional Tables

As is well known today, the limit $\frac{\sin \theta}{\theta}$ as θ approaches 0 is the number 1. The fact that θ in this particular problem is extremely small ("zeroish") means that $\sin \theta$ will differ from it only slightly. In disposing of the problem of finding the sine of a very small angle, many elementary trigonometry textbooks suggest simply using the angle value (in radians) as the answer. Thus, if the angle is 0.00000001 (radians), the sine would be regarded as the same (0.00000001).

Such a difference between an angle and function may prove so slight that conventional tables or readily available values for θ and $\sin \theta$, as illustrated, make the evaluating of $(\theta - \sin \theta)$ enormously difficult. At this point, an infinite series expansion for $\sin \theta$, namely,

$$\sin \theta = \theta - \frac{\theta^3}{3!} + \frac{\theta^5}{5!} - \frac{\theta^7}{7!} + \frac{\theta^9}{9!} - \frac{\theta^{11}}{11!} + \ldots + (-1)^{n-1} \frac{\theta^{2n-1}}{(2n-1)!} + \ldots, (-\infty, \infty),$$

is needed. A computer approach makes the virtually unreachable solution now within our grasp. Note the closeness of the angle and its function for small values of θ in the Maclaurin series above. That is, the statement reduces to $\sin \theta \approx \theta$ if the extremely small terms beyond the first are ignored. The famous, unending series thus permits the finding of a chord which is essentially lost if indeed a sufficient number of terms is considered.

The world of zero offers its many puzzles, fallacies, and paradoxes (Zeno's Paradox, Euler's Doubly Convergent Series, Indeterminate Forms, etc.). Extremely small, virtually negligible quantities likewise offer their diverse challenges. In such a perspective as the latter, the "lost chord" proves an exciting pursuit, both mathematically and historically. Fortunately, the tools of the late twentieth century permit an elegant and detailed numerical disposition far exceeding the computational reach of the mathematical giants of yesteryear. All in all, the "lost chord" provides for an intriguing but speculative entanglement of a Golden Period of Greek Mathematics, the Gaussian Era, and of course, the Age of the Computer. ❑

References

Adler, I. 1957. *The Impossible in Mathematics.* Washington, DC: National Council of Teachers of Mathematics.

Burton, D.M. 1989. *Elementary Number Theory,* Second Edition. Wm. C. Brown Publishers.

Eves, H.W. 1963. *A Survey of Geometry.* Boston: Allyn and Bacon, Inc. Vol. I:225.

——. 1990. *An Introduction to the History of Mathematics.* Philadelphia: Saunders College Publishing.

Francis, R.L. March 1978. A Note on Angle Construction. *The College Mathematics Journal* 16: 75-80.

——. April 1986. Did Gauss Discover That Too? *Mathematics Teacher* 79: 288-293.

——. Spring 1990. Numbers Which Merge, Mix, and Mingle. *Pi Mu Epsilon Journal* 9: 100-105.

Gauss, Carl Friedrich. 1966. *Disquisitiones Arithmeticae.* Arthur A. Clarke, S.J., trans. New Haven: Yale University Press.

Gellert, W., H. Küstner, M. Hellwich, and H. Kästner. 1975. *The VNR Concise Encyclopedia of Mathematics.* New York: Van Nostrand Reinhold Company, 98–101.

Kazarinoff, N.D. 1968. On Who First Proved the Impossibility of Constructing Regular Polygons with Ruler and Compass Alone. *American Mathematical Monthly* 75: 647.

Klein, F.W. Sheppard, P. MacMahon, and L. Mordell. 1962. *Famous Problems and Other Monographs.* New York: Chelsea.

Newson, C.V. and J.F. Randolph. February 1946. Trigonometry without Angles. *Mathematics Teacher* 39: 66-68.

Schaff, William L. 1964. *Carl Friedrich Gauss: Prince of Mathematicians.* New York: Franklin Watts, Inc.

Weisner, L. 1938. *Introduction to the Theory of Equations.* New York: MacMillan.

Appreciation is expressed to W.H. McPherson for the computer verification of certain results of this paper.

The Mathematician of the Millennium

RICHARD L. FRANCIS

Certain intervals on the time line of history demand an assessment. This is true not only in terms of momentous achievement, but also in terms of the singling out of notable men and women. Because of our counting emphasis on powers of ten, significant time periods often take the form of centuries. Reflecting on the subject of the great and the near-great of the nineteenth century was thus a timely task some nine or ten decades ago. Let us now, however, take the process a step further and examine an even broader expanse of years.

Interestingly, we stand at the portals, not only of a new century, but also of a new millennium. The time is thus right for looking back over a vast period of time. As the year 2000 looms on the horizon, we shall turn back in our thinking to the thousand years past and ponder its milestones. As we do, the Magna Carta, the Renaissance, the discovery of America, the Declaration of Independence, and man's setting foot on the moon all clearly stand out as important events. But, in a more narrow setting, what might be said of mathematics across the ten centuries? What were its advancements as we scan the mathematical landscape both near and far, and who were its mathematicians of stature? More precisely, *who was the Mathematician of the Millennium*?

In the world of A.D. 1000, the period of the Dark Ages was drawing to a close. Had a historian of that year bothered to examine the millennium gone by, only by a glance at the Middle Greek Era in western culture would any mathematicians of note be found. Not so today! An avalanche of achievement and a parade of heroes and heroines of mathematics meet the eye of the historically curious.

Out of the massive amounts of mathematics surveyed, where then is one to begin? How might a perspective of the last millennium be gained? A look at 12 distinct periods in the history of mathematics assists. (See **Figure 1**.)

FOOTPRINTS ON THE SANDS OF TIME

Questions of perspective lead immediately to an impressive assortment of names, as well as a careful consideration of the diverse characteristics of mathematical achievement. What features and capabilities might be associated with the mathematically great of the long span of years? It seems fitting to say that the typical mathematical giant would, in some substantial degree, have been

1. A DISCOVERER of powerful methods of resolution, methods that by far transcend the crude approximative and cumbersome techniques of elementary mathematics.

2. A UNIFIER and active seeker-finder of far-reaching concepts, in the sense of integration of ideas, thus identifying and removing many of the superficial compartments of a varied discipline.

3. DARING, both in terms of mainstream mathematics and those branches of mathematics that lead in a pioneering direction.

4. INSIGHTFUL; a poser of problems, and thus a possessor of a grasp of mathematical truth that is truly distinctive.

5. PROLIFIC and hence a highly productive contributor to mankind's storehouse of mathematical knowledge.

6. A UNIVERSALIST and thus a person of great knowledge and vision in the context of the broad and abundant scope of mathematics.

7. INFLUENTIAL; having considerable impact on the mathematics of succeeding generations.

8. INDISPUTABLE in the record; a person whose achievements are not clouded in doubt or cloaked in obscurity.

9. A SURMOUNTER, a barrier-crusher, a mountain-mover, and thus an individual identified with breakthroughs in reference to problems and confinements of a long-standing nature.

10. An ORGANIZER in the sense of an accurate and coherent expression of gained mathematical insights.

It is important for the sake of history that we acknowledge the complexity of relationships. The achievements of the great rest in varying degrees on the accomplishments of predecessors and contemporaries. In some way, we see these notables as holders of the mathematical torch handed to them.

Suppose the traits above are quantified. In the analysis that follows, they will be weighted equally, though this is admittedly debatable. A subtle weighting may nevertheless exist as such items as 1, 2, 3, and 4 all relate to mathematical intuition and power of penetrating judgment. A scale of ten (highest) to zero (lowest) is utilized. The consensus, based on the varied responses, leads to findings not at all surprising.

PARADE OF THE PROMINENT

Only four of the mathematicians of the past millennium received a score of 90 or more in the evaluative scheme. Listed chronologically, they are Isaac Newton, Leonhard Euler, Carl Friedrich Gauss, and David Hilbert. The more detailed assessment appears in **Table 1**.

Obviously, these four mathematicians had much in common. The traits listed confirm this. However, other likenesses bear mentioning. Among these is the fact that all lived to a ripe old age (quite unlike Niels Henrik Abel and Evariste Galois). Enhanced productivity of these conscientious, highly talented people is a clear-cut consequence of such longevity.

All are classified as universalists. In the judgment of some, Newton, Euler, Gauss, and Hilbert were the last of this impressive category of mathematicians. They were likewise teachers, each working on both the theoretical and the applied sides of mathematics. Achievement was of a sustained, lifelong nature.

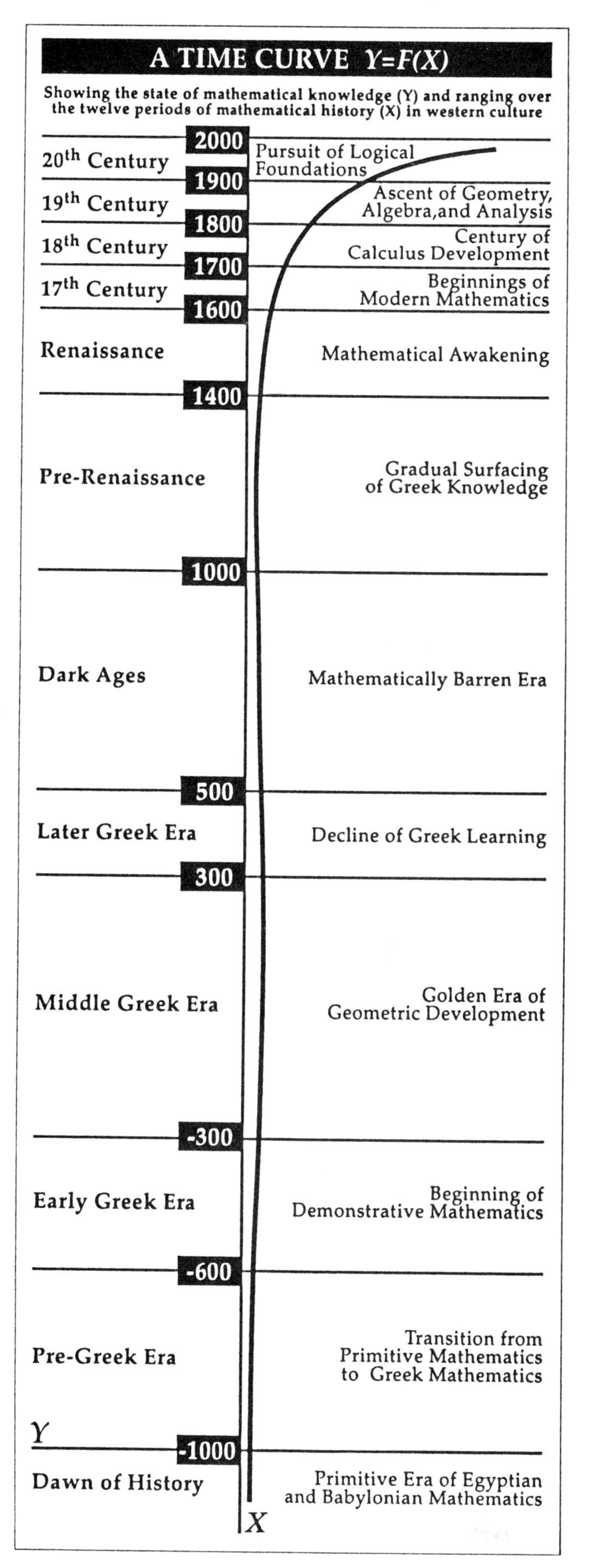

FIGURE 1.

Isaac Newton

From a portrait by Kneller.

Isaac Newton was born December 25, 1642, in Woolsthorpe, England. He died on March 20, 1727. His achievements include

1. the development of the calculus,
2. theoretical pursuit of the general binomial theorem,
3. formulation of the law of universal gravitation,
4. advancements in celestial mechanics and astronomy, and
5. discovery of the three laws of motion that today bear his name.

Leonhard Euler

From a portrait by A. Lorgna.

Leonhard Euler was born April 15, 1707, in Basel, Switzerland. He died on September 18, 1783. His achievements include

1. the early formulation of the basic concepts of network theory and topology,
2. advancements in number theory, including the famous totient or Euler-phi function,
3. notational contributions to accompany varied insights (e, i, $f(x)$, etc.),
4. discovery of the remarkable relationship $e^{\pi i} + 1 = 0$, and
5. discovery of the well-known polyhedral formula $F + V = E + 2$, Euler's constant, and the Euler line of a triangle.

Carl Friedrich Gauss

Portrait origin unknown.

Carl Friedrich Gauss was born April 30, 1777, in Brunswick, Germany. He died on February 23, 1855. His achievements include

1. the discovery of a characterizing theory of regular polygon constructibility,
2. formulation of the law of quadratic reciprocity,
3. discovery (along with others) of the non-Euclidean geometries,
4. advancements in astronomy, and
5. proofs of the Fundamental Theorem of Algebra.

David Hilbert

Courtesy of the Smith Collection, Columbia University

David Hilbert was born January 23, 1862, in Konigsberg, Germany. He died on February 14, 1943. His achievements include

1. the rigorization and further axiomatic development of geometry,
2. establishing the relative consistency of notable geometry types,
3. the extension of Waring's Theorem, whereby a number can be represented as the sum of like powers,
4. the development of the notion of Hilbert space, and
5. the proposal of 23 far-reaching problems in the year 1900, so as to challenge the best efforts of twentieth-century mathematicians.

MATHEMATICIANS OF THE MILLENNIUM	NEWTON	EULER	GAUSS	HILBERT
1. Discoverer	9	8	10	8
2. Unifier	10	8	10	9
3. Daring	10	9	9	9
4. Insightful	10	8	10	9
5. Prolific	10	10	10	9
6. Universalist	10	10	10	10
7. Influential	10	10	10	9
8. Indisputable	9	10	9	10
9. Surmounter	10	9	10	8
10. Organizer	9	9	10	9
TOTAL	**97**	**91**	**98**	**90**

TABLE 1.

Greatness is not always recognized during the lifetimes of those who are later proclaimed so. Not so with Newton, Euler, Gauss, and Hilbert: All were acknowledged as great while they lived. Consider accordingly some of the landmark achievements of these four eminent mathematicians.

It is difficult to visualize the working environment of the mathematicians of the millennium, although all lived in the Modern Era. It is not so difficult, with the mind's eye, to superimpose (incorrectly) the climate of the present on the mathematical world in which they lived. Actually, most carried on in the modest surroundings of poor or cumbersome notation without highly effective calculating devices, and in some state of isolation, considering the slowness of communication. Such a meager background magnifies all the more their achievements.

It has been said that Archimedes would have discovered the calculus if he had had access to a better system of numeration. Such speculation and the playing of the game of "What if?" prove intriguing if applied to mathematicians of the more recent setting. Perhaps Gauss or Hilbert would have proved Fermat's Theorem if either had had a modern day computer available. Perhaps the cardinality of the set of Fermat primes could have been determined. Perhaps ....

The judgment of history hinges on perspective. Unfortunately, such a perspective is difficult if not impossible to obtain as achievers and achievements come close in time to the here-and-now. Consequently, and unfortunately also, mathematicians of the late twentieth century are placed at a disadvantage in the present evaluative scheme. For some, their mathematical quests are just beginning; many are in the midst of a growing achievement.

Now, we return to the original question. Who is, in this account, acclaimed as the Mathematician of the Millennium? My guess would be Carl Friedrich Gauss.

The heritage of this mathematical giant still prompts and directs much of our thinking. Such a heritage may well be described as a wonderland of inviting insights and enticing explorations for the modern day mathematician to ponder and pursue. It is a legacy certain to cross the dividing lines of time and thus inspire mathematicians in the millennium to follow. ❑

References

Buhler, W.K. 1981. *Gauss, A Biographical Study*. New York: Springer-Verlag.

Euler, L. 1984. *Elements of Algebra*. Translated by J. Hewlett, New York: Springer-Verlag.

Eves, H.W. 1990. *An Introduction to the History of Mathematics*. Philadelphia: Saunders College Publishing.

Francis, R.L. 1978. *The Twelve Distinct Eras of Mathematical History in Western Culture*. Fort Wayne: Geyer Instructional Aids Co., Inc.

Gauss, C.F. 1966. *Disquisitiones Arithmeticae*. Arthur A. Clarke, S. J., trans. New Haven: Yale University Press.

Manuel, F.E. 1968. *A Portrait of Isaac Newton*. Cambridge, MA: Harvard University Press.

Olsen, L.M. 1974. *Women in Mathematics*. Cambridge, MA: MIT Press.

Reed, C. 1970. *Hilbert*. New York: Springer-Verlag.

Schaff, W. L. 1964. *Carl Friedrich Gauss: Prince of Mathematicians*. New York: Franklin Watts, Inc.

Turnbull, H.W. 1945. *Mathematical Discoveries of Newton*. Glasgow: Blackie and Sons.

———. 1961. *The Great Mathematicians*. New York: New York University Press.

Imagine That!

A History of Imaginary Numbers

KAREN DOYLE WALTON

In the evolution of mathematics, the invention of new numbers was necessitated by the progress of civilization. The history of $\sqrt{-1}$, the imaginary unit i, follows the logical development of algebraic theory [NCTM 1969]. When trying to express a solution to equations such as $x^2 + 4 = 0$, mathematicians resorted to blind manipulation of symbols, without seriously attempting to understand or interpret their results. The great mathematician Augustine Louis Cauchy was so skeptical of meaningless unintelligible symbols that he condemned roots of negative numbers (called "imaginary numbers") in 1847: "[we discard] the symbolic sign $\sqrt{-1}$, which we repudiate completely, and which we may abandon without regret, because one does not know what this alleged sign signifies, nor what meaning one should attribute to it" [Bell 1945]. However, the end of the nineteenth century saw the acceptance of imaginary numbers by all important investigators due to the irrefutable influence of Carl Friedrich Gauss (1777–1855), "The Prince of Mathematicians"[Cajori 1869].

> "that wonderful creature of an ideal work, almost an amphibian between things that are and things that are not,"

The first trace of the root of a negative number appeared in *Stereometrica* by Heron of Alexandria in about A.D. 50 and took the form $\sqrt{81 - 144}$ [NCTM 1969]. In A.D. 275, Diophantus attempted to compute the sides of a right triangle with perimeter 12 and area 7. His problem required the solution of the equation $336x^2 + 24 = 172x$, which appeared in his *Arithmetica* as the quantity $\sqrt{1{,}849 - 2{,}016}$ (using modern notation) [Smith 1953].

The difficulty of roots of negative numbers was first stated clearly by Mahavira (c. 850) in India who wrote, "As in the nature of things, a negative is not a square, it has no square root." In 1120, the Jewish scholar Abraham bar Chiia expressed difficulty with roots of negatives when he discussed the equations $xy = 48$ and $x + y = 14$. Bhaskara echoed Mahavira in his *Bij Ganita* (1150):

> *the square of an affirmative or of a negative quantity is affirmative; and the square root of an affirmative quantity is two-fold, positive and negative. There is no square-root of a negative quantity: for it is not a square.*
>
> [Smith 1953]

In Europe in the 1400s, Nicolas Chuquet (1484) found that $\sqrt{-a}$ is impossible and Luca Pacioli (1494) stated in his *Suma* that $x^2 + c = bx$ cannot be solved unless $\frac{1}{4}b^2 \leq c$ [Smith 1953].

Although Girolamo Cardano (1545, also known as Jerome Cardan) regarded complex numbers (see **Figure 1**) as "fictitious," he is credited with making progress in solving problems such as the following: "Divide 10 into two parts such that the product ... is 40." Cardano arrived at the "truly sophisticated" solutions $5 + \sqrt{-15}$ and $5 - \sqrt{-15}$, but stated that continuing to work with such quantities would be "as subtle as it would be useless." By 1572, Bombelli had made no improvement on Cardan's theory, but in solving the equation $x^2 + a = 0$, he spoke of the quantities " $+\sqrt{-a}$ " and "$-\sqrt{-a}$" [NCTM 1969].

The first idea of any geometrical interpretation of complex numbers is found in *Algebra* by J. Wallis (1616–1703), an English mathematician and fashionable preacher [Bell 1945, NCTM 1969]. Wallis argued that "when it comes to a Physical Application, [a Negative Quantity] denotes as Real a Quantity as if the Sign were +; but to be interpreted in a contrary sense." Admitting of negative lines and negative areas, he considered a "Negative Plain" with –1600 square perches (160 square perches = 1 English acre) and asked the following question:

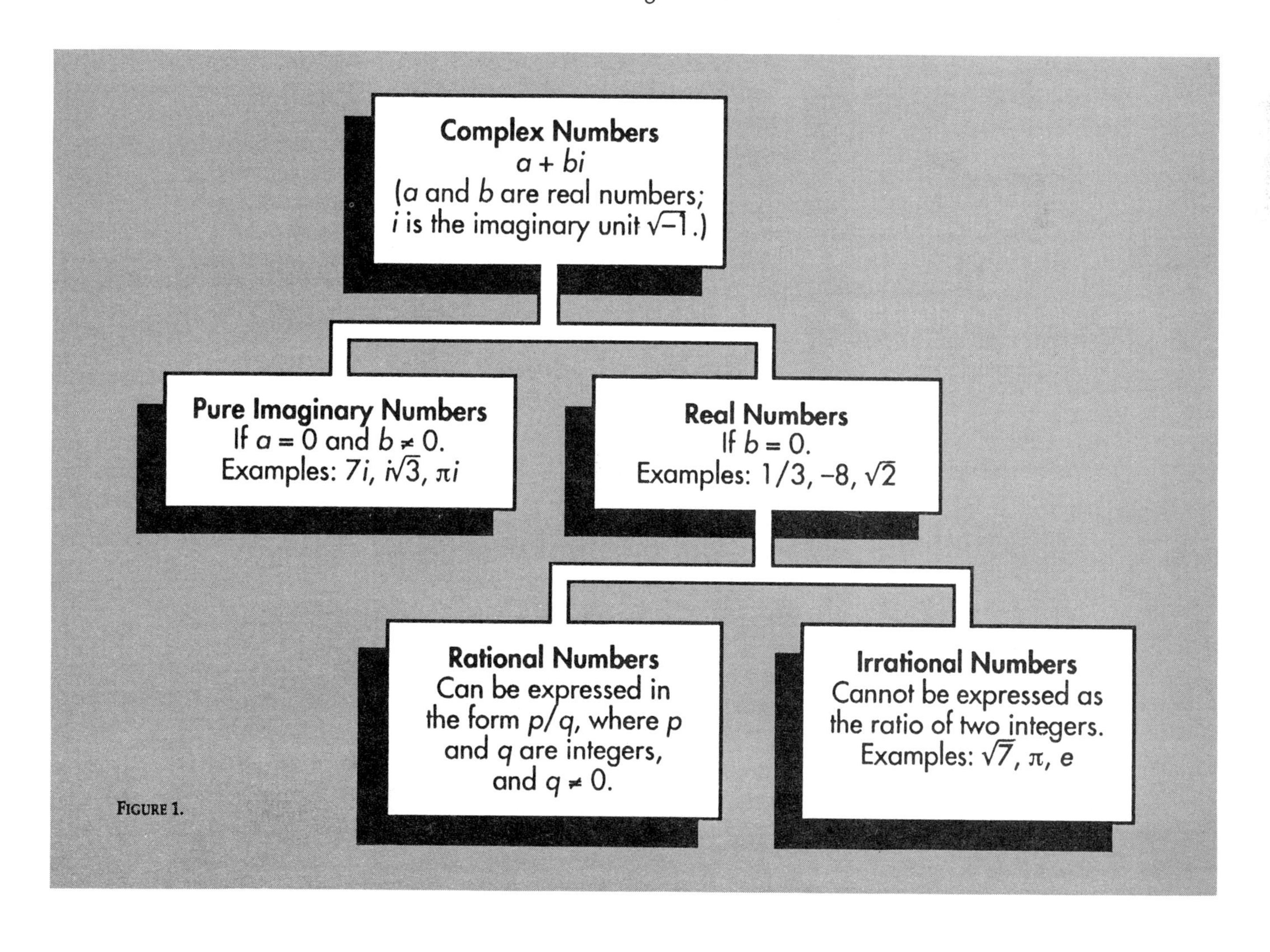

FIGURE 1.

> *What shall this side be? We cannot say it is 40, nor that it is –40. (Because either of these multiplied into itself, will make + 1600; not –1600). Because thus rather, that it is $\sqrt{-1600}$, (the Supposed Root of a Negative Square:) or (which is equivalent thereunto) 10 $\sqrt{-16}$, or 20 $\sqrt{-4}$, or 40 $\sqrt{-1}$.*
>
> [NCTM 1969]

The Norwegian surveyor Caspar Wessel (1745–1818) was the first to make a significant contribution to the understanding of complex numbers through graphical interpretation. Although Wessel presented a completely consistent representation, his presentation to the Royal Danish Academy was published (1799) in a journal not frequently read by mathematicians. It was not until 100 years after his revelation to the Academy that a French translation of the paper appeared. Hence he received only posthumous fame for his work that could have greatly influenced the development of the number system. In 1806, Jean Robert Argand of Geneva published an *Essai* in which he graphed complex numbers on diagrams that represented their formal algebra.

Although Wessel's publication predated Argand's by a century, texts customarily misnamed such graphs as *Armand diagrams* [Bell 1945]. As with non-Euclidean geometry, this was a case of mathematicians working independently of each other, but arriving at similar conclusions.

Other great mathematicians who played roles in the development of complex numbers include Rene Descartes (1637), who originated the terms "real" and "imaginary," Gottfried Wilhelm von Leibniz (1702), who spoke of the imaginary number as "that wonderful creature of an ideal work, almost an amphibian between things that are and things that are not," Leonard Euler (1748), who introduced the letter *i* for $\sqrt{-i}$, and Cauchy (1821), who contributed the terms "conjugate" and "modulus." Despite the efforts of these remarkable mathematicians, it took the awesome influence of Gauss for these numbers, which he named "complex," to gain acceptance by the mathematical world. Gauss found imaginary numbers to be no more objectionable than negative quantities, and unaware of Wessel's work, he made complex numbers seem less mysterious by plotting them on a graph as illustrated in **Figure 2** [Muir 1963].

Gauss's method of graphing complex numbers is presented in elementary algebra textbooks, but the misnomer "imaginary" is too well established to eradicate it from the literature [Bell 1937].

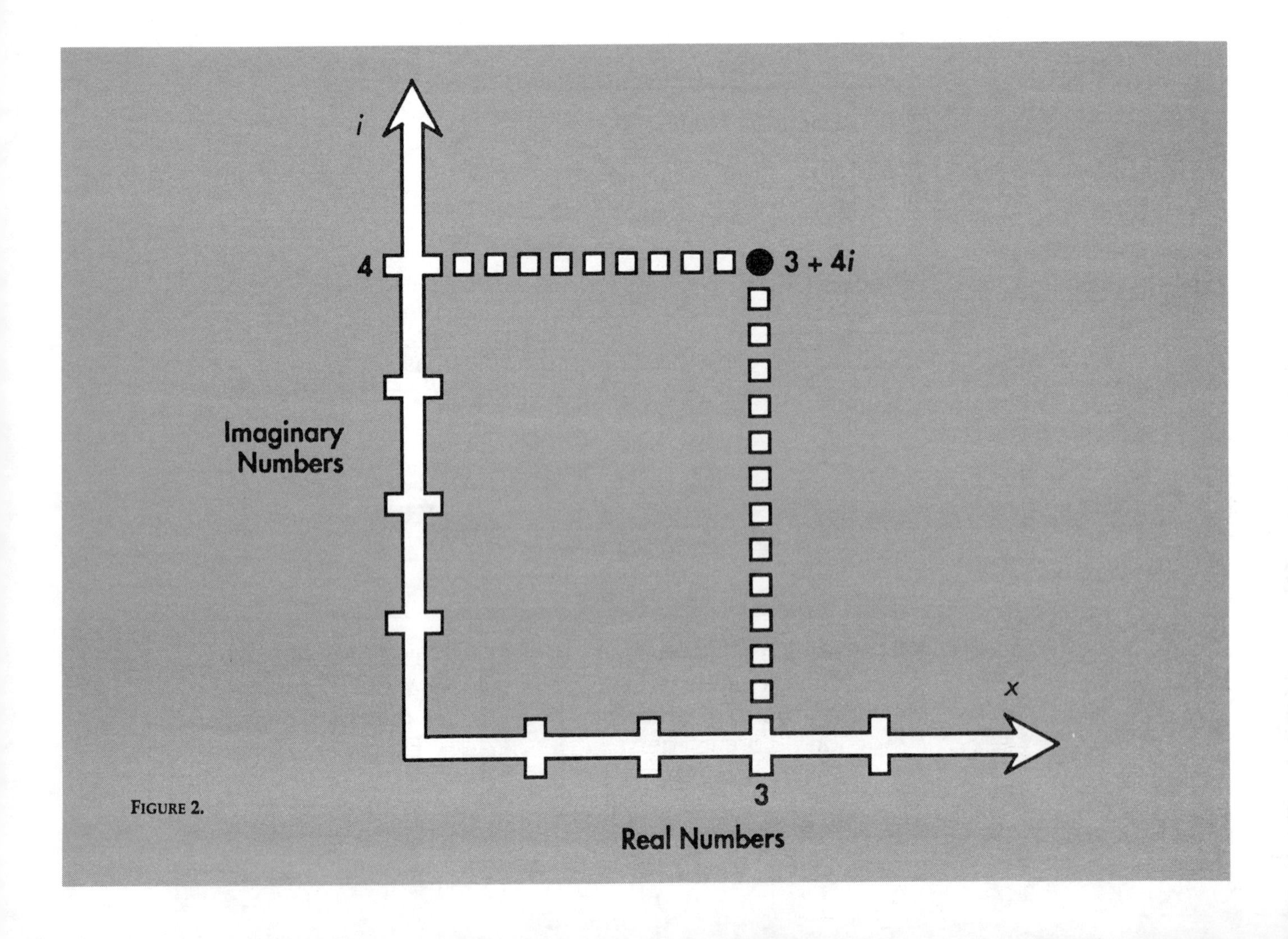

FIGURE 2.

By 1811, Gauss had decided that a solid theory of complex numbers needed a "formal" foundation, meaning the accepted postulates of arithmetic from which the properties of complex numbers could be deduced. Consequently, by 1831 he had invented such a "true metaphysics" of complex numbers, which totally ignored the geometric interpretation. Gauss defined the complex number $a + bi$(a, b are real) as the ordered pair (a, b) on which he defined operations that satisfied the desired postulates. In particular, $(a, b) = (c, d)$ is defined to mean $a = c, b = d$; addition, $(a, b) + (c, d)$, by definition is $(a + c, b + d)$; and multiplication, $(a, b) \times (c, d)$, is $(ac - bd, ad + bc)$.

Six years later in 1837, William Rowan Hamilton presented the same theory, which he discovered independently, to the Royal Irish Academy. Gauss made known his anticipation of Hamilton's work in a letter to his old university friend Wolfgang von Bolyai (1775–1856), to whom Gauss had also revealed via letter his early discovery of non-Euclidean geometry [Bell 1945].

Hence the acceptance of complex numbers affected algebra, analysis, and geometry. In *The Development of Mathematics*, E.T. Bell states that "after about 23 centuries of sightless wandering, arithmeticians and algebraists opened their eyes and saw what Euclid had done: definitions, postulates, deduction, theorems. Then they took a long stride ahead." Although both Gauss and Hamilton applied Euclidean methodology to formulate the theory of complex numbers, as with non-Euclidean geometry, it was the power of Gauss's reputation that assured the credibility of his "true metaphysics" of imaginary numbers.

Students who find it difficult to accept imaginary numbers can be consoled by the fact that for centuries, famous mathematicians were also skeptical. However, after observing the applications of complex numbers in applied mathematics and physics, students are more willing to accept them as an important part of mathematics. ❑

Although both Gauss and Hamilton applied Euclidean methodology to formulate the theory of complex numbers, as with non-Euclidean geometry, it was the power of Gauss's reputation that assured the credibility of his "true metaphysics" of imaginary numbers.

References

Bell, E.T. 1936. *Men of Mathematics*. New York: Simon and Schuster.

———. 1945. *The Development of Mathematics*. New York: McGraw-Hill Book Company, Inc.

Cajori, Florian. 1986. *A History of Elementary Mathematics*. New York: The Macmillan Company.

Muir, Jane. 1963. *Of Men and Numbers*. New York: Dodd, Mead & Company.

National Council of Teachers of Mathematics. 1969. *Historical Topics for the Mathematics Classroom, Thirty-first Yearbook*. Washington, DC: NCTM.

Smith, David E. 1953. *History of Mathematics, Volume II*. Boston: Ginn and Company.

Jewel in the Crown
The Discovery of the Law of Quadratic Reciprocity

RICHARD L. FRANCIS

Rather well known is the story of the heroes of the Persian fable *The Three Princes of Serendip*. Each hero possessed the ability to find extremely valuable things while looking for something else. This desirable gift, called *serendipity*, has made its wonderful appearance in mathematics as well. It is nicely illustrated in a totally unexpected finding by the great mathematician named Carl Friedrich Gauss (1777–1855).

Mathematicians are well aware of the fact that the rational number $1/x$ results in an unending repeating decimal (zeros, of course, may appear). Digits within the repeating block form what is called a *repetend*. For example, $1/7 = 0.142857142857142857...$ or $0.\overline{142857}$. In this case, the repeating block consists of six digits. More interestingly, Gauss, in quest of generalized repetend results, was to find something far better. The discovery was that of the *Law of Quadratic Reciprocity*. It was as if a child, looking for a lost nickel, found unexpectedly a $20 gold piece. Our look at Gauss's serendipitous finding in mathematics has something in common with such a coin search. It requires first a consideration of various mathematical facts and an associated but highly useful symbolism.

A remarkable achievement of the late eighteenth century was the introduction of *congruence notation*, which appears as "≡," into the growing language of number theory symbolism. The congruence symbol is a critical tool with which the mathematician can give representation to a vast assortment of relationships, theorems, and related conjectures. Congruence notation, said to be a "strange equal sign," can be attributed to Gauss, who is credited with introducing it to the mathematical world in his famous 1801 work, *Disquisitiones Arithmeticae*. Gauss's rationale for the shape of the symbol was that it was extremely similar in function to numerical equality. Such a relation of classical number theory, namely, $a \equiv b \pmod{m}$, means that $a - b$ is divisible by m. Somewhat like equality, congruence is reflexive, symmetric, and transitive. Actually, the symbol "≡" was a part of Gauss's explorations and personal writings well prior to its nineteenth century introduction in published form.

However, in spite of the insightful nature of the *Disquisitiones Arithmeticae*, the book's impact on the mathematical community was slow in coming. Virtually three decades were to pass before such notables as Jacobi and Dirichlet brought the publication's deeper content and symbolism, including congruence, to worldwide attention. Its content and field of exploration encompassed far-reaching mathematical relationships. Within this field is what Gauss fittingly called "the gem of arithmetic."

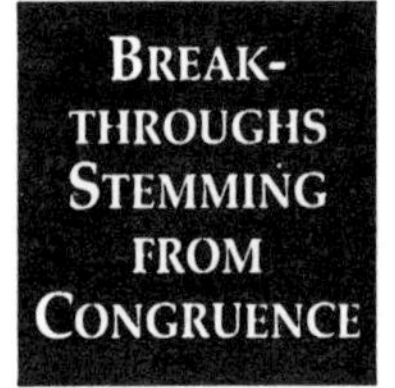

Mathematical notions are of greater importance than the notation by which they are expressed. Though Gauss is credited with this evaluation, one should not routinely minimize the matter of symbolism. Recall that Edward Waring (1734–1798) had made a rather dismal forecast about the unlikely nature of mathematicians' ever proving Wilson's Theorem, considering the inadequacy of notation. However, the theorem is today not only proved, but compactly stated and easily derived in a manner that capitalizes on the concise features of congruence notation. Wilson's Theorem reads,

> An integer n, which is greater than 1, is prime if and only if $(n-1)! \equiv -1 \pmod{n}$.

Note the simplicity of the symbolism; it expresses with minimal notation both a necessary and sufficient condition of primality and opens the door to further inquiry. Waring's dismal prediction thus ended on a bright note.

Finite systems, including rings and fields, Diophantine equations, equivalence classes, the calendar, and clock or modular arithmetic all fit beautifully into this general frame of congruence. With the advent of the new notation, other well-known facts from decades past took on a new look. Fermat's Theorem, which revealed that certain powers upon division by primes always yielded remainders of 1, became $a^{p-1} \equiv 1 \pmod{p}$, if prime p is not a divisor of a. In a similar fashion, Euler's Theorem became $a^{\phi(m)} \equiv 1 \pmod{m}$. Here, another notational item emerges, namely, the famous *totient* or *Euler phi-function*.

Theorems invariably raise the question concerning the validity of their converses. Fermat's Theorem works unfailingly for all primes. However, could the congruence result now provide a test for primality? That is, if n satisfies $a^{n-1} \equiv 1 \pmod{n}$, and a is not divisible by n, must n necessarily be a prime? Oriental mathematicians had pursued this question over 2500 years ago, though the problem was posed essentially in a rhetorical manner (not in a way that utilized the new notation). Counter-examples were sought, yet the search proved frustratingly difficult.

However, in 1819, it was discovered that $2^{340} \equiv 1 \pmod{341}$. As $341 = (11)(31)$ is thus a composite number satisfying the Fermat condition, a new class of numbers came to light. They are called *pseudo-primes*. Such numbers are today known to constitute an infinite set. Still the original, longstanding quest culminated in an accurate conclusion only by a simple but symbolic tool called congruence. Further explorations and discoveries have similarly been expedited by the easily applied rules and nicely manipulated symbolism it affords. A notable example is that of Gauss's analysis of the Easter date pattern; it inherently involved the concepts of congruence arithmetic. (See box, page 84.)

The rules that govern the solving of congruences of the form $ax \equiv b \pmod{m}$ were quickly formalized by Gauss. They paralleled closely the fundamental rules of equation solving with the first major departure being that of a cancellation or division law. In reference to a given modulus m, solutions were generally restricted to the set consisting of $0, 1, 2, 3, \ldots, m-1$. Such a collection is today called *the set of least non-negative residues*. A close connection was seen to exist with the ancient techniques of applying the Euclidean algorithm and solving linear Diophantine equations.

A congruence parallel can likewise be drawn to the long-term development of a theory of equations. Solving quadratic equations, decidedly more difficult than their linear counterparts, led historically to construction concerns, more complicated techniques, and (ultimately) to a broader class of quantities called *complex numbers*. Much the same was realized in moving from the simpler setting of linear congruences to the deeper area of development in the case for higher degree congruences. In this latter exploration, Gauss was again to play an outstanding role.

Most are familiar with the process of completing the square and the quick introduction it provides of fractions and radicals in solving $ax^2 + bx + c = 0$ (where $a \neq 0$). As the domain of classical number theory is the set of integers, another approach emerges in solving the congruence counterpart, $x^2 + bx + c \equiv 0 \pmod{p}$, in which a is not divisible by prime p. This follows as fractions and radicals are not ideal forms in the congruence setting. First consider this alternate approach in solving the general quadratic equation. Such an endeavor paves the way to solving quadratic congruences.

$$ax^2 + bx + c = 0$$
$$4a^2x^2 + 4abx = -4ac \quad \text{[multiplying by } 4a]$$
$$4a^2x^2 + 4abx + b^2 = b^2 - 4ac \quad \text{[adding } b^2]$$
$$(2ax + b)^2 = b^2 - 4ac.$$

In the field of complex numbers, the work is quickly completed. By extracting the square root of each member, x is thus found.

In solving quadratic congruences, the question centers around whether or not there is an integer $2ax + b$ (call this y) such that $y^2 \equiv (b^2 - 4ac) \pmod{p}$. More concisely, if $b^2 - 4ac$ is called D, the problem concerning the solvability of $y^2 \equiv D \pmod{p}$ becomes critical. Again, any quadratic congruence $ax^2 + bx + c \equiv 0 \pmod{p}$ can be transformed to $y^2 \equiv D \pmod{p}$ by the sequence of steps shown above. It will have exactly two solutions or no solutions. We will return to this critical matter shortly, but first, here's an example of the technique. Application will be made of the fact that the modulus may be added or subtracted any number of times to or from either member of the congruence statement.

$$x^2 + 7x - 1 \equiv 0 \pmod{11}$$
$$4x^2 + 28x \equiv 4 \pmod{11}$$
$$\text{[multiplying by } 4a]$$
$$4x^2 + 28x + 49 \equiv 53 \pmod{11}$$
$$\text{[adding } b^2]$$
$$(2x + 7)^2 \equiv 53 \pmod{11}$$
$$(2x + 7)^2 \equiv 9 \pmod{11}$$
$$\text{[subtracting (4)(11) from 53]}$$
$$2x + 7 \equiv 3 \pmod{11}, \text{ or}$$
$$2x + 7 \equiv -3 \pmod{11}.$$

Therefore, x may be assigned the values -2 or -5, or equivalently, x is 9 or x is 6 by adding the modulus 11 to each number.

The most subtle part of the solution above appeared in line 5. Is it actually possible to solve a congruence of the form $y^2 \equiv D \pmod{p}$? Sometimes the answer is yes. But not always! It is here that some of the more challenging aspects of congruence theory begin to surface.

The Easter Date Pattern

The selection of various days of celebration for church holidays have been made over the centuries by different civilizations and with different methods. Many times, holiday dates depended on the cycles of the moon, so they depended upon the calendar in use at that period in time. Mathematicians were often called upon to determine holiday dates, and Gauss was no exception.

In the year A.D. 325, the Council of Nice decreed that Easter would fall on the first Sunday after the full moon that happens on or closest to March 21, the vernal equinox. If the full moon occurs on a Sunday, Easter would be the following Sunday. In modern times, Easter will always fall between March 22 and April 25, as determined from the Book of Common Prayer or some other church document. Without giving the details of how or why this works, the following shows the formula discovered by Gauss in 1800 to determine on what date Easter will fall.

1. Select appropriate numbers under the headings m and n from the table below:

Years	m	n
1582–1699	22	2
1700–1799	23	3
1800–1899	23	4
1900–2000	24	5

   For example, for the year 1992, $m = 24$ and $n = 5$.

2. Divide the number of the year by 4, and call the remainder a:

   $1992/4 = 498$, and a remainder of 0, or $a = 0$.

3. Divide the number of the year by 7, and call the remainder b:

   $1992/7 = 284$, R (remainder) 4, or $b = 4$.

4. Divide the number of the year by 19, and call the remainder c:

   $1992/19 = 104$ R16, or $c = 16$.

5. Divide $(19c + m)$ by 30, and call the remainder d:

   $((19)(16) + 24)/30 = 328/30 = 10$ R28, or $d = 28$.

6. Divide $(2a + 4b + 6d + n)$ by 7, and call the remainder e:

   $(2(0) + 4(4) + 6(28) + 5)/7 = 27$ R0, or $e = 0$.

7. Then Easter is either the $(22 + d + e)$th day of March, or the $(d + e - 9)$th day of April; since $(22 + d + e) = 50$, in this year 1992, Easter must fall on the $(d + e - 9)$th day of April, or April 19.

   Now, you try the year 1999!

The Law of Quadratic Reciprocity

Should the congruence $y^2 \equiv D \pmod{p}$ have a solution (where p is an odd prime and D is not a multiple of p), then D is called a *quadratic residue* of p. Otherwise, D is a quadratic non-residue of p. It can be shown that p has as many residues as non-residues. The great Swiss mathematician, Leonhard Euler, had pursued this matter but did not make use of congruence symbolism. In the notation of the present, he established what is called *Euler Criterion*, namely, that D is a quadratic residue of p if and only if $D^{(p-1)/2} \equiv 1 \pmod{p}$. By Fermat's Theorem, we know that only a 1 or a –1 will appear to the right of the congruence symbol. In the event of a –1 result, D is a quadratic non-residue of p. Adrien-Marie Legendre further supplemented the relationship by introducing the notation that $(D/p) = 1$ if D is a quadratic residue of p, and $(D/p) = -1$ if D is a quadratic non-residue of p.

Both Euler and Legendre were fascinated by a certain outgrowth of these pursuits, namely, a pairing of related quadratic congruences so that each congruence complemented the other. Nevertheless, it remained for Gauss to give a concise (symbolic) formulation of the problem as well as to supply a rigorous proof of its validity. Euler, shortly prior to his death in 1783, had conjectured what is today called the *Law of Quadratic Reciprocity*. His explorations were essentially of an inductive nature. In just two years (1785), Legendre was to construct an inadequate proof. Legendre's pitfall involved the use of assumptions that were as hard to prove as the quadratic reciprocity law itself.

The lifespans of Euler and Gauss barely overlapped. However, Gauss and Legendre (Gauss's elder by 25 years), were involved in some controversy. The controversy concerned questions of priority of discovery and to whom should credit for a first, rigorous proof be given. Such a challenging and remarkable property concerns distinct odd primes, say p and q, and reads as shown below.

Law of Quadratic Reciprocity
(The Gem of Arithmetic)

$$(p/q)\,(q/p) = (-1)^{\frac{(p-1)}{2}\frac{(q-1)}{2}}.$$

The symbols (p/q) and (q/p) are the Legendre symbols mentioned earlier. Conciseness of the relationship is apparent. Yet, the Law of Quadratic Reciprocity is far-reaching in consequence. It enables the mathematician to find (p/q) if (q/p) is known, or vice versa.

It appears that Gauss did not know of the prior work of Euler and Legendre in this area. Accordingly, his finding of the law was most likely one of independent discovery. The serendipitous finding was in the year 1795 when Gauss was but eighteen years of age. In the following year, he produced a fully rigorous proof of the Law of Quadratic Reciprocity. It was a mathematical first. (Over the years, Gauss would give eight different proofs of the now famous law. This overkill approach is reminiscent of Gauss's four distinct proofs of the Fundamental Theorem of Algebra.) Actually, the year 1796 was a monumental one in terms of mathematical advancement. Not only was the Law of Quadratic Reciprocity resolved, so too was the removing of a construction limitation of the ancients, namely, Gauss's construction of the regular seventeen-sided polygon. The year 1996, looming on the calendar, is a two-hundredth anniversary of both discoveries.

Gauss proudly asserted that mathematics is the queen of the sciences. His further remarks identify number theory as its crowning achievement. But he says even more. The Law of Quadratic Reciprocity is boldly labeled the "gem of arithmetic," the jewel in the crown.

Gauss later endeavored to extend his explorations to include higher order congruences. Though the domain of classical number theory is that of the integers, a general disposition of cubic and quartic congruences required the introduction, as discovered by Gauss, of numbers called *Gaussian integers*. Such integers are those numbers of the form $a + bi$, where a and b are ordinary integers. The Gaussian plane, which is a complex number plane, graphically reinforces this long ago encounter.

The time period of the late eighteenth century and the early nineteenth century was a highly productive one, arithmetically speaking. Superficial aspects of the discipline and accompanying conjectures were, more and more, being examined cautiously so as to give rise to a degree of rigorization. It was a progressive form of rigorization, one that today characterizes all of mathematics. In the unfolding story, the Law of Quadratic Reciprocity, a gem of arithmetic, even the jewel in the crown, played a shining role. ❑

References

Burton, D. M. 1976. *Elementary Number Theory*. Boston: Allyn and Bacon.

Francis, R.L. 1988. *A Mathematical Look at the Calendar*. Lexington, MA: COMAP, Inc.

Gauss, C.F. 1966. *Disquisitiones Arithmeticae*. Arthur A. Clarke, S. J., trans. New Haven: Yale University Press.

Newman J.R. 1956. *The World of Mathematics*. Vol. 1. New York: Simon and Schuster.

Schaff, W.L. 1964. *Immortals of Science: Carl Friedrich Gauss, Prince of Mathematicians*. New York: Franklin Watts, Inc.

The Three Tools of Euclidean Construction

RICHARD L. FRANCIS

Certain constructions identified in elementary geometry are known to be impossible if the allowable instruments are restricted to the unmarked straightedge and the compass (the *Euclidean instruments*). Among these are the three famous problems of antiquity, namely, *trisecting the general angle, duplicating a cube*, and *squaring a circle*. These famous problems are reasonable outgrowths of such basic constructions as bisecting an angle, duplicating a square, and squaring a polygon.

Geometry students today are well-versed in these basic constructions and accordingly find it challenging to extend their efforts in the direction of the classical problems. This fascination likely continues because of some unawareness or persistent skepticism of the fact that each construction is known to be impossible. Actually, the three problems were negatively resolved in the nineteenth century, a time of considerable activity in the developing of a theory of equations and transcendental classifications. The well-known instruments in all these constructions are the tools that produce the most fundamental of curves, that is, the line and the circle.

Rarely mentioned is a third restriction, which is, in fact, a vital part of all classical Euclidean constructions. It is a limitation which, if ignored, makes all of the above named constructions possible. Such a restriction comes more clearly into focus in the steps of the processes that follow.

Look at **Figure 1**. Suppose angle θ is to be trisected using only the unmarked straightedge and compass. We will now capitalize on the fact that angle bisection and repeated bisections are available techniques; this implies the validity of angle quadrisection (double bisection). Quadrisecting angle θ in a repeated, unending manner gives rise to the series

$$\frac{\theta}{4}+\frac{\theta}{16}+\frac{\theta}{64}+\frac{\theta}{256}+\ldots.$$

But this is an infinite geometric series that converges. Its first term a is $\theta/4$ and its ration r is $1/4$. Using the formula $S_\infty = a/(1-r)$, such a series is seen to converge to

FIGURE 1.

$$\frac{\frac{\theta}{4}}{1-\frac{1}{4}}$$

or $\theta/3$. The angle θ is thus trisected using only the Euclidean tools.

Of course, the construction, as is clearly evident, required infinitely many steps. This is the feature which pinpoints the third but subtle restriction referred to above. Very simply, *all Euclidean constructions must be performed in a finite number of steps*. The angle trisection here considered, so seemingly acceptable at first glance, is now seen to fall short.

If repeated constructions involving an infinity of steps are allowed, an angle can be divided by compass and straightedge into n equal parts for any positive integer n whatever. Consider $n = 11$, for example. As $\theta/12$ can be obtained by quadrisecting θ and then trisecting such a quadrisection (using the infinite process above), the geometric series

$$\frac{\theta}{12}+\frac{\theta}{144}+\frac{\theta}{1728}+\ldots$$

results and yields $\theta/11$ for its convergent sum. A more general treatment for n-secting an angle θ takes the form of

$$\sum_{i=1}^{\infty}\frac{\theta}{(n+1)^i}.$$

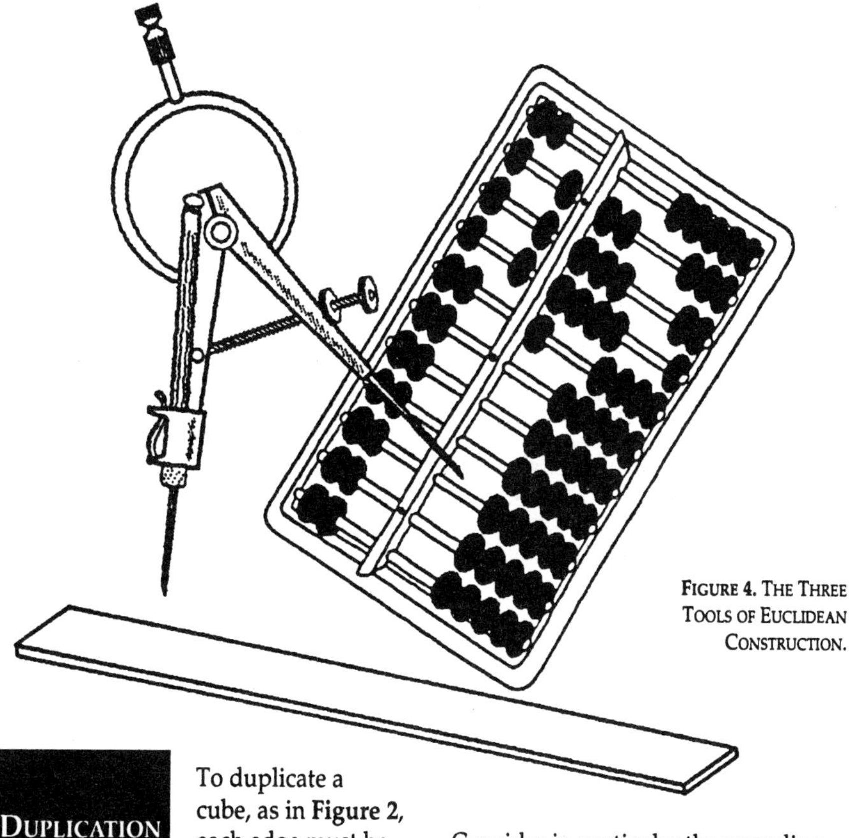

FIGURE 4. THE THREE TOOLS OF EUCLIDEAN CONSTRUCTION.

DUPLICATION OF A CUBE

To duplicate a cube, as in **Figure 2**, each edge must be multiplied by the cube root of 2. But, can the cube root of 2 be constructed with the unmarked straightedge and compass? Recall that the square root of 2 is constructible, as are the fourth root, the eighth root, the sixteenth root, etc. These are actually repeated square roots. Also recall that the multiplication of line segment lengths is a valid Euclidean construction.

FIGURE 2.

Consider in particular the unending product

$$2^{\frac{1}{4}} \cdot 2^{\frac{1}{16}} \cdot 2^{\frac{1}{64}} \cdot 2^{\frac{1}{256}} \cdot \ldots$$

in which each factor after the first is the fourth root of the preceding. Adding exponents, an infinite geometric series, which is also convergent, appears.

Accordingly, the product becomes

$$2^{\frac{1}{4}+\frac{1}{16}+\frac{1}{64}+\frac{1}{256}+\ldots}$$

or $2^{1/3}$. The cube root of 2 clearly emerges.

Availability of $\sqrt[3]{2}$ makes $e\sqrt[3]{2}$ also constructible. The original cube of edge e is resultingly duplicated. As before, the construction, performed entirely with the Euclidean instruments, has required an infinity of steps and must be rejected.

If infinitely many steps in the construction process are permissible, the cube root of 3 would also prove constructible. This would follow in the same manner as above simply by replacing the base 2 by a base of 3. Hence, a cube may be triplicated. More generally, the cube root of an arbitrary positive integer n can be constructed by a similar modification. This allows for the multiplication of the volume of a cube by any such number n whatever.

It is possible, with the Euclidean tools, to construct a square having the same area as any polygon. That is, any polygon can be squared. In order to square a circle, begin with a polygon inscribed in the circle. In particular, consider a circle O_r, which circumscribes an equilateral triangle. If the number of sides of the triangle is doubled, still preserving the regular and inscribed features, the resulting hexagon will have an area that better approximates the area of the circle. (See **Figure 3**.) Suppose now that the doubling process, subject to the inscribed and regular conditions, continues endlessly. Then the area approximation of polygon to circle becomes ever closer. Very simply, if A_{R_n} denotes the area of the regular inscribed polygon of n sides, we know that

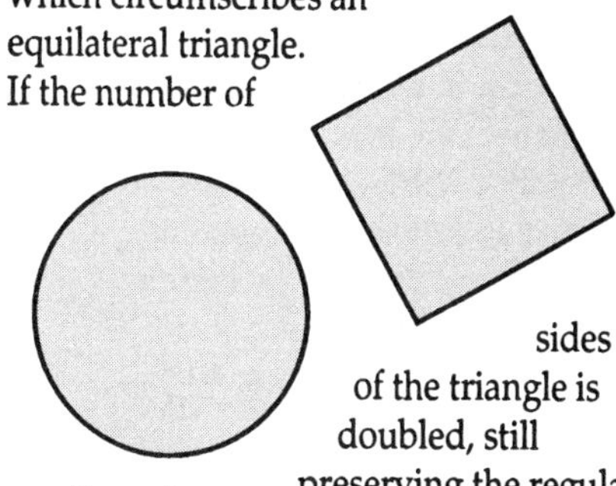

FIGURE 3.

$$\lim_{n \to \infty} A_{R_n} = \pi r^2.$$

But any polygon can be squared. Hence, the polygons of the limiting process above permit us ultimately to square the circle. We see again however a construction which necessitated an infinite number of steps. As before, we reject it.

A careful consideration of infinitely many segments also proves revealing. Since any line segment (including the unit segment) can be multisected with the Euclidean tools, and, as circle squaring requires the construction of π, proceed by first utilizing the well-known Gregory series for representing π. More precisely, consider

$$\frac{\pi}{4} = 1 - \frac{1}{3} + \frac{1}{5} - \frac{1}{7} + \frac{1}{9} - \frac{1}{11} + \ldots.$$

Suppose that infinitely many steps are allowed. Since each of the rational numbers in the series representation of π is constructible, so does the transcendental number π itself prove constructible. A square could thus be constructed having the same area as the given circle.

All of the constructions above were performed with the Euclidean instruments. However, each required an infinite number of steps. Such procedures are identified as *asymptotic Euclidean constructions*. They provide, in a rather straightforward manner, challenging twists on the famous problems of antiquity.

The line and the circle restriction of a Euclidean construction has, in an analytic setting, resulted in the various proofs of impossibility which we know well today. Of course, the instrument counterparts to these curves, the straightedge and the compass, identify the various construction attempts in a visible, tangible, and highly conspicuous manner. However, what instrument from the ancient world best symbolizes the third restriction, that is, the finite one? Likely, it should be some simple but finite counting device, one which is so characteristically a part of a remote time setting. Appropriately, we will now, for ornamental purposes and in a symbolic manner, choose the *abacus* as a third but subtle instrument of a classical Euclidean construction. It serves as an embellishment that is a consistent *counting* reminder of the precluding of an infinity of steps.

It was established by Lorenzo Mascheroni that all classical Euclidean constructions of a point-by-point nature could be performed with the compass alone. This discovery was included in Mascheroni's 1797 publication *Geometria del Compasso*. Actually, the fact that the compass alone was sufficient was known a century earlier; it had appeared in a publication by Joseph Mohr. However, this previous publication was lost for many years. It was found only in the year 1928. Jean-Victor Poncelet and Jacob Steiner were to discover that the straightedge alone was likewise sufficient, provided at least one circular arc with its center was given. Hence, there is some possibility of restricting these two basic instruments of construction. In no case, however, can the finite condition be compromised if in fact the objective is that of a classical Euclidean construction. Such a critical condition is symbolized by the abacus.

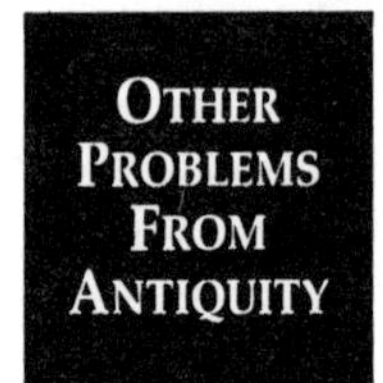

Consider also another famous problem bequeathed to mathematicians by the ancient world: the problem of regular polygon constructibility.

Though the ancient Greeks were successful in constructing polygons of three, four, five, and six sides (as in **Figure 5**), they could not construct the regular heptagon (seven-sided polygon). Gauss (1777–1855) was to conjecture in 1801 in his *Disquisitiones Arithmeticae* that such a Euclidean construction is impossible. Complete verification of the conjecture is the work of Pierre Laurent Wantzel (1814–1848) in the year 1837.

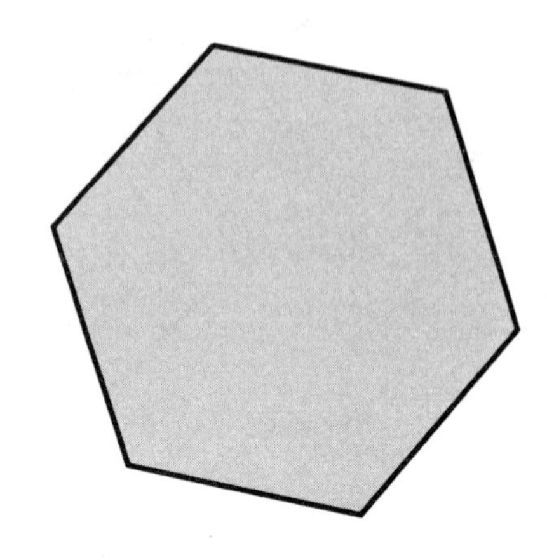

Figure 5.

However, suppose (again) that infinitely many steps are permissible. As an angle can be divided by eight (successive bisection), then any angle θ may be used to generate the expression

$$\frac{\theta}{8} + \frac{\theta}{64} + \frac{\theta}{512} + \frac{\theta}{4096} + \ldots.$$

This is an infinite geometric series whose first term a is $\theta/8$ and whose ratio r is $1/8$. Use of the now familiar formula $S_\infty = a/(1 - r)$ results in the angle measure $(\theta/8)/(1 - (1/8))$ or $\theta/7$. By letting $\theta = 360°$, the central angle of the regular heptagon, i. e., $360°/7°$, has thus been constructed. The reader may wish to consider still other regular polygon construction types. Though the Euclidean constructibility criterion rules out, for example, the regular nonagon (nine-sided polygon), a single trisection of the 120° angle (asymptotically of course) gives rise to the desired 40° angle. This is the central angle of a regular nonagon.

Note: the series above may be modified so as to construct $2^{1/7}$. That is, in an asymptotic setting, $\sqrt[7]{2}$ is a constructible number. More generally, asymptotic methods allow for the construction of the *n*th root of any whole number.

A *Fermat prime* is a prime which is 1 more than a power of 2. A *Mersenne prime* is a prime 1 less than a power of 2. The only prime in both categories is the number 3. Gauss, Wantzel, and others established that any regular polygon having a prime number of sides is constructible if and only if that number